SpringerBriefs in Mathematics

SpringerBriefs in Mathematics showcases expositions in all areas of mathematics and applied mathematics. Manuscripts presenting new results or a single new result in a classical field, new field, or an emerging topic, applications, or bridges between new results and already published works, are encouraged. The series is intended for mathematicians and applied mathematicians.

For further volumes:
www.springer.com/series/10030

Tatsien Li · Yongji Tan · Zhijie Cai · Wei Chen ·
Jingnong Wang

Mathematical Model
of Spontaneous Potential
Well-Logging and
Its Numerical Solutions

 Springer

Tatsien Li
School of Mathematical Sciences
Fudan University
Shanghai, People's Republic of China

Yongji Tan
School of Mathematical Sciences
Fudan University
Shanghai, People's Republic of China

Zhijie Cai
School of Mathematical Sciences
Fudan University
Shanghai, People's Republic of China

Wei Chen
Shanghai Lixin University of Commerce
Shanghai, People's Republic of China

Jingnong Wang
Technology Center
China Petroleum Logging Co. Ltd
Xi'an, People's Republic of China

ISSN 2191-8198 ISSN 2191-8201 (electronic)
ISBN 978-3-642-41424-4 ISBN 978-3-642-41425-1 (eBook)
DOI 10.1007/978-3-642-41425-1
Springer Heidelberg New York Dordrecht London

Mathematics Subject Classification: 35J25, 46E35, 65N30, 86A20

Printed on acid-free paper

Springer is part of Springer Science+Business Media (www.springer.com)

Preface

The spontaneous potential well-logging is one of the most common and useful well-logging techniques in petroleum exploitation. Since the earth layer can be regarded as a piecewise uniform medium, and due to electrochemical and other factors, there exists a jump of potential (spontaneous potential difference) on each interface of different layers. The corresponding mathematical model should be the boundary value problem of quasi-harmonic partial differential equations with inhomogeneous interface conditions. In axi-symmetric situation, at the crossing point of multiple interfaces, the compatible condition is usually violated so that it is not possible to get a solution to the boundary value problem in the sense of piecewise H^1 space, and we are obliged to consider the solution in piecewise $W^{1,p}$ space ($1 < p < 2$). This is an important feature of spontaneous potential well-logging, which naturally makes the necessity of introducing Sobolev spaces with fractional power, and seriously increases the difficulty of proving the well-posedness and proposing numerical solution schemes. In this book, in the axi-symmetric situation we demonstrate the well-posedness of the corresponding mathematical model, and develop three efficient schemes of numerical solution to meet the need of practical computation, the soul of which is to eliminate the non-compatibility (singularity) first, and then to put it into the common solution process of the Finite Element Method. Besides, according to the characteristic of the problem, we set up certain results on the limiting behavior of solutions, which notably simplify the computation in many cases. In the meantime, we offer a series of numerical examples according to the practical petroleum well-logging to reveal some important properties of spontaneous potential well-logging and validate some of the theoretic conclusions.

There are very few reference material about the mathematical model and solution technique for the spontaneous potential well-logging. This book is a systematical summing up of the research results for many years of our research team. Peng Yuejun, Li Hailong and Zhou Yi also made important contributions to this research project, and we would like to express our sincere thanks to them. We owe our appreciation to Hubei Jianghan Well-Logging Institute who initially posed this project and stimulated us to persist in the corresponding research in long term. We

also grateful to Technology Center of China Petroleum Logging Co. Ltd. of China
National Petroleum Cooperation (CNPC) for the cooperation with us in the recent
years, which pushed the research forward deeper and more fruitful.

Shanghai, People's Republic of China   Tatsien Li
Yongji Tan
Zhijie Cai
Wei Chen
Jingnong Wang

Contents

Chapter 1
Modeling

In petroleum exploitation one often uses various methods of well-logging, among which the spontaneous potential (SP) well-logging is one of the most common and useful techniques. Due to electrochemical and other factors, the positive ions and negative ions have different migration velocities in the layer, and shale particles usually attract positive ions, a stable potential difference called the SP difference can be formed on every interface of different layers, and an electric field in the earth layer is then produced by these SP differences. We call this electric field as the SP electric field to distinguish it from the artificial electric field produced by the power supply. The distribution of SP electric field is closely related to the lithology. Especially, in the sand/shale section it can show the permeable formation by a significant abnormal change of the SP curve. Therefore, through the study of SP electric field the lithology of the wall around the well can be revealed. This technique is named the spontaneous potential well-logging. With methodological simplicity and high practicality the spontaneous potential well-logging has become one of the basic and indispensable methods of studying the lithology and reservoir properties, obtaining the well-logging parameters and other geological applications [1, 2, 6, 7].

To do the spontaneous potential well-logging, a measuring electrode is placed within the well bore and its potential is measured. As the measuring electrode is raised continuously, by measuring the potential on it a corresponding spontaneous potential curve (SP curve) is obtained. The SP curve varies with the variation of the formation and the location of the permeable formation can be clearly shown.

As shown in Fig. 1.1, we suppose that the formation is symmetric about the well axis and the central level, where Ω_1 is the mud area with resistivity ρ_1; Ω_2 is the enclosing rock with resistivity ρ_2, Ω_3 and Ω_4 are two sections of the objective layer. In Ω_3 the resistivity is changed since the filtering of mud into the objective layer, then Ω_3 is called the invaded zone with resistivity ρ_3, and Ω_4 is the original objective layer with resistivity ρ_4.

The SP electric field obeys essentially the law of a steady one. Let $u = u(x, y, z)$ be the potential function of the field. We will first introduce the differential equation and the boundary conditions for it.

T. Li et al., *Mathematical Model of Spontaneous Potential Well-Logging and Its Numerical Solutions*, SpringerBriefs in Mathematics, DOI 10.1007/978-3-642-41425-1_1, © The Author(s) 2014

Fig. 1.1 Model of the formation

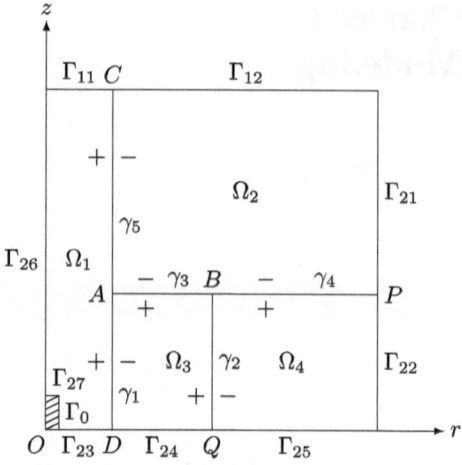

In an electric field caused by steady currents, the current density function j satisfies the continuity equation

$$\operatorname{div} j = 0. \tag{1.1}$$

By Ohm's law in differential form

$$j = \frac{E}{\rho}, \tag{1.2}$$

where ρ is the resistivity and E is the intensity of the electric field with

$$E = -\operatorname{grad} u, \tag{1.3}$$

in each subdomain with continuous resistivity, we have

$$\operatorname{div}\left(\frac{1}{\rho}\operatorname{grad} u\right) = 0, \tag{1.4}$$

i.e., the potential function u in the formation satisfies the following quasi-harmonic equation [3–5]:

$$\frac{\partial}{\partial x}\left(\frac{1}{\rho}\frac{\partial u}{\partial x}\right) + \frac{\partial}{\partial y}\left(\frac{1}{\rho}\frac{\partial u}{\partial y}\right) + \frac{\partial}{\partial z}\left(\frac{1}{\rho}\frac{\partial u}{\partial z}\right) = 0 \quad \text{in } \Omega_i \ (i = 1, 2, 3, 4), \tag{1.5}$$

where ρ is the resistivity of the formation, $\rho = \rho_i$ in each subdomain Ω_i ($i = 1, 2, 3, 4$). By denoting $\sigma = \frac{1}{\rho}$ as the conductivity, $\sigma = \sigma_i = \frac{1}{\rho_i}$ in each subdomain Ω_i ($i = 1, 2, 3, 4$). In this book, we always suppose that $\sigma_i > 0$ ($i = 1, 2, 3, 4$).

Under the hypothesis of axi-symmetry, (1.1) can be written as the axi-symmetric form as follows:

$$\frac{\partial}{\partial r}\left(\sigma r \frac{\partial u}{\partial r}\right) + \frac{\partial}{\partial z}\left(\sigma r \frac{\partial u}{\partial z}\right) = 0 \quad \text{in } \Omega_i \ (i = 1, 2, 3, 4), \tag{1.6}$$

where $u = u(r, z)$ with $r = \sqrt{x^2 + y^2}$.

We take the potential value on the earth surface Γ_{12} to be zero as the reference potential:

$$u|_{\Gamma_{12}} = 0. \tag{1.7}$$

On the plane of symmetry, symmetric axis, infinite boundary and the insulating surface of the electrode (namely, on $\bigcup_{k=1}^{7} \Gamma_{2k} \overset{\text{def.}}{=} \Gamma_2$), since there is no current passing through, the boundary condition will be

$$\sigma r \frac{\partial u}{\partial n}\bigg|_{\Gamma_2} = 0, \tag{1.8}$$

where n is the unit outward norm vector of Γ_2 and $\dfrac{\partial u}{\partial n}$ is the outward normal derivative of u.

As to the measuring electrode which is mainly used to measure the potential value on its electrode surface, since there is no current emission from it, and it is made of good conductor, so that the potential on its surface should be equal to a constant (to be determined), on the electrode surface (denoted by Γ_0) u should satisfy a equi-potential surface boundary condition as follows:

$$u|_{\Gamma_0} = U \quad \text{(a constant to be determined)}, \tag{1.9}$$

$$\int_{\Gamma_0} \sigma r \frac{\partial u}{\partial n} dS = 0. \tag{1.10}$$

On the interfaces of different domains (denoted by $\gamma_k, k = 1, \ldots, 5$), by the current continuity, the following interface condition should be satisfied:

$$\left(\sigma r \frac{\partial u}{\partial n}\right)^{+}\bigg|_{\gamma_k} = \left(\sigma r \frac{\partial u}{\partial n}\right)^{-}\bigg|_{\gamma_k} \quad (k = 1, \ldots, 5), \tag{1.11}$$

moreover, due to the spontaneous potential difference on the interface of different media, we have

$$\left(u^{+} - u^{-}\right)\big|_{\gamma_k} = E_k \quad (k = 1, \ldots, 5), \tag{1.12}$$

where E_k is the spontaneous potential difference on the interface $\gamma_k \ (k = 1, \ldots, 5)$, which is supposed to be a constant in the real applications, "+" and "−" stand for the values of a function on the both sides of the interface, respectively, and the unit normal vector n of the interface is set to have the same orientation on both sides.

By noticing the compatibility with the interface condition (1.12), the potential function u on Γ_{11}, a part of the earth surface, should be taken a constant value as

$$u|_{\Gamma_{11}} = E_5. \tag{1.13}$$

Thus in the axi-symmetric case $u = u(r, z)$ satisfies a boundary value problem (1.6)–(1.13).

We make a transformation

$$\tilde{u} = u - u_0, \tag{1.14}$$

where u_0 is a function of piecewise constant as follows:

$$u_0 = \begin{cases} E_5 & \text{in } \Omega_1, \\ 0 & \text{in } \Omega_2, \\ E_5 - E_1 & \text{in } \Omega_3, \\ E_5 - E_1 - E_2 & \text{in } \Omega_4. \end{cases} \tag{1.15}$$

It is easy to see that $\tilde{u}$ satisfies a similar boundary value problem as u, however, the interface condition (1.12) changes to (for convenience, here we still use u to represent $\tilde{u}$)

$$\left. \left(u^+ - u^- \right) \right|_{\gamma_k} = F_k \quad (k = 1, \ldots, 5), \tag{1.16}$$

where

$$F_1 = F_2 = F_5 = 0, \qquad F_3 = E_1 + E_3 - E_5 \stackrel{\text{def.}}{=} \Delta_A,$$
$$F_4 = E_1 + E_2 + E_4 - E_5 \stackrel{\text{def.}}{=} \Delta_A + \Delta_B. \tag{1.17}$$

It means that there is no spontaneous potential difference on the vertical interfaces, correspondingly, boundary conditions (1.7) and (1.13) on $\Gamma_1 = \Gamma_{11} \cup \Gamma_{12}$ can be uniformly written as

$$u = 0 \quad \text{on } \Gamma_1. \tag{1.18}$$

Obviously, the transformed problem (1.6), (1.18), (1.8)–(1.11) and (1.16)–(1.17) is simpler than the original problem. Back to the original problem, it is easy to obtain the following

Theorem 1.1 *For any given geometric structure and formation resistivity, the values of spontaneous potential function on the electrode, on the well-axis and in domains Ω_1 and Ω_2 depend only on the constant E_5, $E_1 + E_3$ and $E_1 + E_2 + E_4$, but not on the 5 constants E_k ($k = 1, \ldots, 5$) independently.*

This theorem will greatly reduce the cost of producing the corresponding well-logging interpretation chart.

So, through the transformation (1.14), we obtain the mathematical model for the spontaneous potential well-logging, that is a boundary value problem of quasi-harmonic equation (1.6) with the corresponding boundary conditions (1.18), (1.8)–(1.10) and the interface conditions (1.11) and (1.16). For simplicity of statement, this model is denoted by (SP) later on.

We may consider a more general case (which will be used in the discussion below) that the spontaneous potential difference E_k on the interface γ_k may not be a constant but a function denoted by $E_k(s)$ (for $k = 1, \ldots, 5$), where s is the arc length of γ_k calculating from the starting points A (for $k = 1, 3, 5$) or B (for $k = 2, 4$), respectively, and $E_k(s)$ is supposed to be Hölder continuous. To ensure the compatibility of the interface condition (1.12) with the boundary condition (1.8), we suppose that

$$E_1'(D) = E_2'(Q) = E_4'(P) = 0. \tag{1.19}$$

In addition, to ensure the compatibility of the interface condition (1.12) with the boundary condition (1.7), boundary condition (1.13) should be

$$u|_{\Gamma_{11}} = E_5(C). \tag{1.20}$$

Through a simple translation $\tilde{u} = u - v$, where

$$v = \begin{cases} E_5(C) & \text{in } \Omega_1, \\ 0 & \text{in } \Omega_i \ (i = 2, 3, 4), \end{cases} \tag{1.21}$$

we are also able to write the boundary conditions (1.7) and (1.20) uniformly as (1.18), and then we may suppose that

$$E_5(C) = 0. \tag{1.22}$$

Later on in the solution of spontaneous potential problem, for $k = 1, \ldots, 5$, E_k is a general function only in a neighborhood of A and B, and still a constant elsewhere, then (1.19) will be satisfied naturally.

References

1. Chu, Z., Huang, R., Gao, J., Xiao, L.: Methods and Principle of Geophysical Well-Loggings (Volume I). Oil Industry Press, Beijing (2007) (in Chinese)
2. Heart, J.R., Nelson, P.H.: Well Logging for Physical Properties. McGraw-Hill, New York (1985)
3. Li, T., et al.: Applications of the Finite Element Method in Electric Well-Loggings. Oil Industry Press, Beijing (1980) (in Chinese)
4. Li, T., Tan, Y., Peng, Y.: Mathematical model and method for the spontaneous potential well-logging. Eur. J. Appl. Math. **5**, 123–139 (1994)
5. Li, T., et al.: Boundary Value Problems with Equivalued Surface and Resistivity Well-Logging. Pitman Research Notes in Mathematics Series, vol. 382. Chapman & Hall/CRC, Boca Raton (1998)
6. Pan, H., Ma, H., Niu, Y.: Geophysical Well-Loggings and Borehole Geophysical Prospecting. Science Press, Beijing (2009) (in Chinese)
7. Zhang, G.: Electric Well-Loggings. Petroleum University Press, Beijing (1996) (in Chinese)

Chapter 2
Properties of Solutions

2.1 Space of Solutions

To solve the model (SP), we will consider the equivalent variational problem. For this purpose we first introduce several function sets.

For any given p $(1 < p < \infty)$, we introduce the set

$$V_p = \{v \mid v \in W_*^{1,p}(\Omega_i) \ (i = 1, 2, 3, 4), \ (v^+ - v^-)\big|_{\gamma_k} = E_k(s) \ (k = 1, \ldots, 5),$$

$$v\big|_{\Gamma_1} = 0, \ v\big|_{\Gamma_0} = \text{constant}\}, \tag{2.1}$$

where $W_*^{1,p}$ $(1 < p < \infty)$ stands for the following weighted Sobolev space:

$$W_*^{1,p} = \{v \mid r^{1/p} v \in L^p, r^{1/p} \nabla v \in L^p\} \tag{2.2}$$

endowed with the norm

$$\|v\|_{W_*^{1,p}} = \left(\|r^{1/p} v\|_{L^p}^p + \|r^{1/p} \nabla v\|_{L^p}^p\right)^{\frac{1}{p}}, \tag{2.3}$$

where $\nabla v = \left(\frac{\partial v}{\partial r}, \frac{\partial v}{\partial z}\right)$. As $p = 2$, we denote $W_*^{1,p} = H_*^1$ (see [1]). Then the variational problem equivalent to the model (SP) is to find $u \in V_p$ such that

$$\sum_{i=1}^4 \iint_{\Omega_i} \sigma_i \left(\frac{\partial u}{\partial r} \frac{\partial \phi}{\partial r} + \frac{\partial u}{\partial z} \frac{\partial \phi}{\partial z}\right) r \, dr \, dz = 0, \quad \forall \phi \in V_{p'}^0, \tag{2.4}$$

where

$$V_{p'}^0 = \{v \mid v \in W_*^{1,p'}(\Omega_i) \ (i = 1, 2, 3, 4), \ (v^+ - v^-)\big|_{\gamma_k} = 0 \ (k = 1, \ldots, 5),$$

$$v\big|_{\Gamma_1} = 0, \ v\big|_{\Gamma_0} = \text{constant}\},$$

$$= \{v \mid v \in W_*^{1,p'}(\Omega), \ v\big|_{\Gamma_1} = 0, \ v\big|_{\Gamma_0} = \text{constant}\}, \tag{2.5}$$

in which $\Omega = \bigcup_{i=1}^4 \Omega_i$ and p' is the dual number of p, i.e., $\frac{1}{p} + \frac{1}{p'} = 1$.

T. Li et al., *Mathematical Model of Spontaneous Potential Well-Logging and Its Numerical Solutions*, SpringerBriefs in Mathematics, DOI 10.1007/978-3-642-41425-1_2, © The Author(s) 2014

If the algebraic sum of the spontaneous differences $E_k(s)$ $(k = 1, \ldots, 5)$ on the interfaces are zero at the crossing points A and B, i.e.,

$$
\begin{cases}
\Delta_A \stackrel{\text{def.}}{=} E_1(A) + E_3(A) - E_5(A) = 0, \\
\Delta_B \stackrel{\text{def.}}{=} E_2(B) - E_3(B) + E_4(B) = 0,
\end{cases}
\tag{2.6}
$$

the spontaneous potential differences are said to satisfy the compatible condition [6]. In this case we can solve the problem (SP) in piecewise H_*^1 space and use the finite element method to get its numerical solution.

However, in general the spontaneous potential differences on the interfaces do not satisfy the compatible condition (2.6), the corresponding potential function in the electric field has singularity and no longer belongs to piecewise H_*^1 space. It causes an essential difficulty not only in theoretical analysis but also in numerical solution since the finite element method can not be directly used to numerically solving the problem in this case.

Now we first consider the set, to which the solution should belong [6, 7].

Lemma 2.1 *For any given $p \geq 2$, let $E_k(s) \in C^\alpha(\gamma_k)$ $(k = 1, \ldots, 5)$, where $\alpha > 1 - \frac{1}{p}$. Then $V_p \neq \emptyset$ (i.e., V_p is not empty) if and only if the compatible condition (2.6) is valid. Moreover, as (2.6) holds, there exists a $w \in V_p$ such that*

$$
\sum_{i=1}^{4} \|w\|_{W_*^{1,p}(\Omega_i)} \leq C(p) \sum_{k=1}^{5} \|E_k\|_{W^{1-1/p,p}(\gamma_k)},
\tag{2.7}
$$

here and hereafter $C(p)$ stands for a positive constant depending only on p.

Proof Without loss of generality, we only prove this lemma in the case $p = 2$. The proof in the general case is similar.

Sufficiency: If (2.6) holds, by the inverse trace theorem on an angular domain [4], there exists a $H_*^1(\Omega_1)$ function w_1 satisfying the following boundary condition:

$$
w_1 =
\begin{cases}
E_1(s) + E_3(\frac{l_3}{l_1}s) & \text{on } \gamma_1^+, \\
0 & \text{on } \Gamma_{11}, \\
E_5(s) & \text{on } \gamma_5^+
\end{cases}
\tag{2.8}
$$

and

$$
\|w_1\|_{H_*^1(\Omega_1)} \leq C \sum_{k=1}^{5} \|E_k\|_{H^{1/2}(\gamma_k)},
\tag{2.9}
$$

where l_k is the length of γ_k $(k = 1, \ldots, 5)$, and by (1.18), $E_5(l_5) = 0$.

Similarly, there exist $w_3 \in H_*^1(\Omega_3)$ and $w_4 \in H_*^1(\Omega_4)$ satisfying the following boundary conditions:

$$
w_3 = \begin{cases} E_3(\frac{l_3}{l_1}s) & \text{on } \gamma_1^-, \\ E_3(s) & \text{on } \gamma_3^+, \\ E_2(s) + E_4(\frac{l_4}{l_2}s) & \text{on } \gamma_2^+, \end{cases} \tag{2.10}
$$

$$
w_4 = \begin{cases} E_4(\frac{l_4}{l_2}s) & \text{on } \gamma_2^-, \\ E_4(s) & \text{on } \gamma_4^+, \end{cases} \tag{2.11}
$$

respectively, and

$$
\|w_3\|_{H_*^1(\Omega_3)}, \|w_4\|_{H_*^1(\Omega_4)} \leq C \sum_{k=1}^{5} \|E_k\|_{H^{1/2}(\gamma_k)}. \tag{2.12}
$$

Denoting

$$
w = \begin{cases} w_1 & \text{in } \Omega_1, \\ 0 & \text{in } \Omega_2, \\ w_3 & \text{in } \Omega_3, \\ w_4 & \text{in } \Omega_4, \end{cases} \tag{2.13}
$$

it is evident that $w \in V_2$, and by (2.9) and (2.12) we know that (2.7) is valid.

Necessity: If (2.6) is not valid, for example, $\Delta_A \neq 0$, by the result of above sufficiency, there exists a piecewise H_*^1 function w such that

$$
w^+ - w^- = \begin{cases} E_1(s) - \Delta_A & \text{on } \gamma_1, \\ E_2(s) & \text{on } \gamma_2, \\ E_3(s) & \text{on } \gamma_3, \\ E_4(s) - \Delta_B & \text{on } \gamma_4, \\ E_5(s) & \text{on } \gamma_5. \end{cases} \tag{2.14}
$$

If $V_2 \neq \emptyset$, then there exists a piecewise H_*^1 function v such that $(v^+ - v^-)|_{\gamma_k} = E_k(s)$ $(k = 1, \ldots, 5)$. Let $u = v - w$. u is also a piecewise H_*^1 function, and

$$
u^+ - u^- = \begin{cases} \Delta_A & \text{on } \gamma_1, \\ 0 & \text{on } \gamma_2, \\ 0 & \text{on } \gamma_3, \\ \Delta_B & \text{on } \gamma_4, \\ 0 & \text{on } \gamma_5. \end{cases} \tag{2.15}
$$

Fig. 2.1 Figure for (2.17)

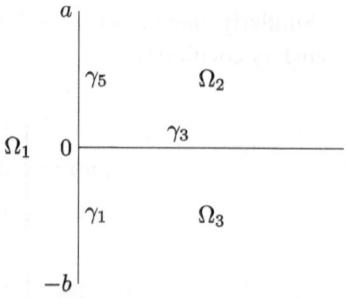

Denoting $u = u_i$ in Ω_i ($i = 1, 2, 3, 4$), since $u_2|_{\gamma_3^-} = u_3|_{\gamma_3^+}$, we have

$$\tilde{u} = \begin{cases} u_2 & \text{in } \Omega_2 \\ u_3 & \text{in } \Omega_3 \end{cases} \in H_*^1(\Omega_2 \cup \Omega_3), \tag{2.16}$$

therefore by the trace theorem, $u_1|_\gamma - \tilde{u}|_\gamma \in H^{1/2}(\gamma)$, where $\gamma = \gamma_1 \cup \gamma_5$.
  On the other hand, by (2.15) we have

$$(u_1 - \tilde{u})|_\gamma = \begin{cases} \Delta_A & \text{on } \gamma_1 \\ 0 & \text{on } \gamma_5 \end{cases} \overset{\text{def.}}{=} f \quad \text{(see Fig. 2.1)}, \tag{2.17}$$

then, by the equivalent norm in $H^{1/2}$ [1], we get

$$\|f\|_{H^{1/2}(\gamma)}^2 = \int_{-b}^a |f(x)|^2 dx + \int_{-b}^a \int_{-b}^a \left| \frac{f(x) - f(y)}{x - y} \right|^2 dxdy$$

$$= \Delta_A^2 b + 2 \int_0^a \int_{-b}^0 \frac{\Delta_A^2}{(x-y)^2} dxdy$$

$$= +\infty.$$

It leads to a contradiction, therefore V_2 must be empty: $V_2 = \emptyset$. □

Lemma 2.2 *For any given* p *($1 < p < 2$), assume that* $E_k(s) \in C^\alpha(\gamma_k)$ *($k = 1, \ldots, 5$), where* $\alpha > 1 - \frac{1}{p}$, *then* $V_p \neq \emptyset$, *and there exists* $w \in V_p$ *such that*

$$\sum_{i=1}^4 \|w\|_{W_*^{1,p}(\Omega_i)} \leq C(p) \sum_{k=1}^5 \|E_k\|_{W^{1-1/p,p}(\gamma_k)}. \tag{2.18}$$

Proof We still use the method of piecewise construction in the proof of the necessity in Lemma 2.1. By the inverse trace theorem it is easy to see that in order to prove Lemma 2.2, it is enough to show $f \in W^{1-1/p,p}(\gamma)$, where f is defined by (2.17).

It can be verified by direct calculation as follows:

$$\|f\|_{W^{1-1/p,p}(\gamma)}^p = \int_{-b}^a |f(x)|^p dx + \int_{-b}^a \int_{-b}^a \frac{|f(x) - f(y)|^p}{|x-y|^p} dx dy$$

$$= \Delta_A^p b + 2 \int_0^a \int_{-b}^0 \frac{\Delta_A^p}{(x-y)^p} dx dy$$

$$< +\infty.$$

The first equality above comes from the equivalent norm of $W^{1-1/p,p}$ [1].   □

2.2 The Existence of Piecewise H_*^1 Solution as the Compatible Condition Is Satisfied

Suppose that the compatible condition (2.6) is satisfied for the spontaneous potential differences on the interfaces, and denote the solution of variational problem (2.4) in V_2 by u. By Lemma 2.1, $V_2 \neq \emptyset$. We take arbitrarily $w \in V_2$, which satisfies (2.7). Denoting $v = u - w \in V_2^0$, then for any given $\phi \in V_2^0$, we have

$$\sum_{i=1}^4 \iint_{\Omega_i} \sigma_i \left(\frac{\partial v}{\partial r} \frac{\partial \phi}{\partial r} + \frac{\partial v}{\partial z} \frac{\partial \phi}{\partial z} \right) r dr dz = - \sum_{i=1}^4 \iint_{\Omega_i} \sigma_i \left(\frac{\partial w}{\partial r} \frac{\partial \phi}{\partial r} + \frac{\partial w}{\partial z} \frac{\partial \phi}{\partial z} \right) r dr dz,$$

$$(2.19)$$

i.e.,

$$\iint_\Omega \sigma \left(\frac{\partial v}{\partial r} \frac{\partial \phi}{\partial r} + \frac{\partial v}{\partial z} \frac{\partial \phi}{\partial z} \right) r dr dz = \iint_\Omega \left(f_1 \frac{\partial \phi}{\partial r} + f_2 \frac{\partial \phi}{\partial z} \right) r dr dz, \qquad (2.20)$$

where $\sigma = \sigma_i$, $f = (f_1, f_2) = \sigma_i \nabla w$ in Ω_i $(i = 1, 2, 3, 4)$. Obviously, $f \in (L_*^2(\Omega))^2$.

By Lax-Milgram theorem [2, 3], the variational problem (2.20) admits a unique solution $v \in V_2^0$, then the variational problem (2.4) admits a unique solution $u = v + w \in V_2$.

2.3 Piecewise $W_*^{1,p}$ Solution and Its Estimation as the Compatible Condition Is Not Satisfied

If the compatible condition (2.6) is not satisfied, by Lemma 2.1, $V_2 = \emptyset$, so the variational problem (2.4) does not admit a piecewise H_*^1 solution. However, by Lemma 2.2, for any given p $(1 < p < 2)$, $V_p \neq \emptyset$, then we can consider the existence of the piecewise $W_*^{1,p}$ solution for the variational problem (2.4) (see [5, 7]). Thus, to solve the (SP) problem, generally speaking, we are obliged to use Sobolev space with fractional power.

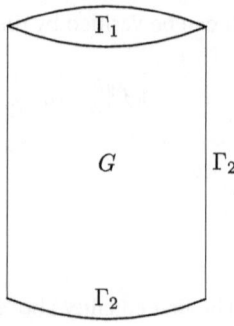

Theorem 2.1 *There exists an $\varepsilon_0 > 0$ suitably small, such that for any given p
$(2 - \varepsilon_0 < p \leq 2)$ and $f \in L_*^p(\Omega)$, there exists a unique solution $v \in V_p^0$ to the variational problem (2.20), and*

$$\|v\|_{W_*^{1,p}(\Omega)} \leq C(p)\|f\|_{L_*^p(\Omega)}. \tag{2.21}$$

Noting that the problem under consideration is reduced from a 3-D problem according to the axi-symmetry, in order to prove Theorem 2.1 we go back to the 3-D
problem and to investigate the following more general situation. Let G be a cylinder
in $\mathbb{R}^3$, Γ_1 be its upper bottom and Γ_2 be its lower bottom and its lateral boundary
(see Fig. 2.2). Denoting $U^p = \{u \mid u \in W^{1,p}(G), u|_{\Gamma_1} = 0\}$, $a(x) \in L^\infty(G)$, with
$0 < a_{\min} \leq a(x) \leq a_{\max} < \infty$ almost everywhere in G.

Theorem 2.2 *There exists an $\varepsilon_0 > 0$ depending only on $a_{\min}$ and $a_{\max}$, such that
for any given p $(2 - \varepsilon_0 < p \leq 2)$ and any given $f \in (L^p(G))^3$, there exists a unique
solution $u \in U^p$ which satisfies*

$$\iiint_G a(x)\nabla u \cdot \nabla\phi \, dx dy dz = \iiint_G f \cdot \nabla\phi \, dx dy dz, \quad \forall \phi \in U^{p'}, \tag{2.22}$$

where p' is the dual number of p, moreover,

$$\|u\|_{W^{1,p}(G)} \leq C\|f\|_{L^p(G)}, \tag{2.23}$$

where C is a positive constant depending only on p, $a_{\min}$ and $a_{\max}$.

To prove Theorem 2.2, we need the following lemmas.

Lemma 2.3 *Let $f \in (C_0^\infty(G))^3$. Then for any given p $(1 < p \leq 2)$, problem*

$$\begin{cases} \Delta u = \operatorname{div} f, \\ u|_{\Gamma_1} = 0, \\ \dfrac{\partial u}{\partial n}\bigg|_{\Gamma_2} = 0 \end{cases} \tag{2.24}$$

admits a unique solution $u \in U^p$, and

$$\|u\|_{W^{1,p}(G)} \leq C(p)\|f\|_{L^p(G)}, \tag{2.25}$$

where $C(p)$ is a positive constant depending on p.

Proof See [2]. □

Noticing that problem (2.24) can be written into the corresponding variational form

$$\iiint_G \nabla u \cdot \nabla \phi \, dx \, dy \, dz = \iiint_G f \cdot \nabla \phi \, dx \, dy \, dz, \quad \forall \phi \in U^{p'}, \tag{2.26}$$

we have

Lemma 2.4 *For any given p ($1 < p \leq 2$) and any given $f \in (L^p(G))^3$, there exists a unique $u \in U^p$ satisfying (2.26) and*

$$\|\nabla u\|_{L^p(G)} \leq C(p)\|f\|_{L^p(G)}. \tag{2.27}$$

Proof Noticing that $C_0^\infty(G)$ is dense in $L^p(G)$, the desired conclusion follows immediately from Lemma 2.3. □

Lemma 2.5 *For any given p ($1 < p \leq 2$), let $C_*(p)$ be the best constant which makes (2.27) valid, i.e.,*

$$C_*(p) = \inf_{\|f\|_{L^p(G)}=1} \|\nabla u\|_{L^p(G)}. \tag{2.28}$$

Then we have

(i) $C_*(2) = 1$;
(ii) $C_*(p)$ is a continuous function of p.

Proof Especially taking $f = \nabla u$ in (2.27), it is easy to see that

$$C_*(p) \geq 1, \quad \forall p \in (1, 2]. \tag{2.29}$$

As $p = 2$, taking $\phi = u$ in (2.26), we obtain

$$\|\nabla u\|_{L^2(G)} \leq \|f\|_{L^2(G)}.$$

Therefore

$$C_*(2) \leq 1. \tag{2.30}$$

By (2.29) and (2.30) we obtain (i) immediately.

To prove (ii), let $d(\frac{1}{p}) = C_*(p)$. We only need to prove that d is a continuous function of p. By

$$\|\nabla u\|_{L^{p_1}(G)} \le d\left(\frac{1}{p_1}\right)\|f\|_{L^{p_1}(G)},$$

$$\|\nabla u\|_{L^{p_2}(G)} \le d\left(\frac{1}{p_2}\right)\|f\|_{L^{p_2}(G)},$$

and noticing $T : f \to u$ is a linear operator, the interpolation theorem leads to

$$\|\nabla u\|_{L^r(G)} \le \left(d\left(\frac{1}{p_1}\right)\right)^{\theta}\left(d\left(\frac{1}{p_2}\right)\right)^{1-\theta}\|f\|_{L^r(G)}, \tag{2.31}$$

where

$$\frac{1}{r} = \frac{\theta}{p_1} + \frac{1-\theta}{p_2}, \quad \forall 0 \le \theta \le 1.$$

Then we have

$$d\left(\frac{\theta}{p_1} + \frac{1-\theta}{p_2}\right) \le \left(d\left(\frac{1}{p_1}\right)\right)^{\theta}\left(d\left(\frac{1}{p_2}\right)\right)^{1-\theta}, \tag{2.32}$$

namely $\ln d$ is a convex function. Since any convex function must be continuous, so $\ln d$ is continuous, i.e., d is continuous. $\qquad\square$

Proof of Theorem 2.2 We first show that for any solution u of (2.22), (2.23) holds.
Let $m = \frac{a_{\max} - a_{\min}}{a_{\max}}$. It is evident that $0 < m < 1$. By (2.22) we have

$$\iiint_G \nabla u \cdot \nabla \phi dx dy dz = \frac{1}{a_{\max}} \iiint_G f \cdot \nabla \phi dx dy dz$$

$$+ \iiint_G \frac{a_{\max} - a(x)}{a_{\max}} \nabla u \cdot \nabla \phi dx dy dz. \tag{2.33}$$

By Lemma 2.4, it is easy to see that

$$\|\nabla u\|_{L^p(G)} \le C_*(p)\left(\frac{1}{a_{\max}}\|f\|_{L^p(G)} + m\|\nabla u\|_{L^p(G)}\right). \tag{2.34}$$

By Lemma 2.5, as p is sufficiently close to 2, we have

$$C_*(p)m < \frac{m+1}{2} < 1. \tag{2.35}$$

Then by (2.34) we obtain

$$\|\nabla u\|_{L^p(G)} \le \frac{2C_*(p)}{a_{\max}(1-m)}\|f\|_{L^p(G)}, \tag{2.36}$$

therefore (2.23) is valid.

By (2.23) we get immediately the uniqueness of the solution of problem (2.22).
Now we prove the existence of the solution of problem (2.22).

Let $a_\delta = J_\delta * a$, where J_δ is the mollifier, "$*$" stands for the convolution. It is easy to see that $a_\delta \in C^\infty(\overline{G})$, $\sup a_\delta \leq a_{max}$, $\inf a_\delta \geq a_{min}$, and as $\delta \to 0$, a_δ converges to a strongly in $L^{p'}(G)$, where p' is the dual number of p. For any fixed $a_\delta \in C^\infty(\overline{G})$, there exists a solution $u_\delta \in U^p$ which satisfies [2]

$$\iiint_G a_\delta(x)\nabla u_\delta \cdot \nabla\phi dxdydz = \iiint_G f \cdot \nabla\phi dxdydz, \quad \forall \phi \in U_{p'}, \qquad (2.37)$$

and (2.23) is valid. By (2.37), we especially have

$$\iiint_G a_\delta(x)\nabla u_\delta \cdot \nabla\phi dxdydz = \iiint_G f \cdot \nabla\phi dxdydz, \quad \forall \phi \in U_0^\infty, \qquad (2.38)$$

where $U_0^\infty = \{u \mid u \in C^\infty(G), u|_{\Gamma_1} = 0\}$. By (2.23), $\{u_\delta\}$ is uniformly bounded with respect to δ in $W^{1,p}(G)$, then there exists a subsequence $\{u_{\delta_k}\}$ which converges weakly in $W^{1,p}(G)$: $u_{\delta_k} \rightharpoonup u$ as $k \to +\infty$. Taking $\delta = \delta_k$ in (2.38) and setting $k \to \infty$, we obtain

$$\iiint_G a(x)\nabla u \cdot \nabla\phi dxdydz = \iiint_G f \cdot \nabla\phi dxdydz, \quad \forall \phi \in C_0^\infty(G),$$

then

$$\iiint_G a(x)\nabla u \cdot \nabla\phi dxdydz = \iiint_G f \cdot \nabla\phi dxdydz, \quad \forall \phi \in U_{p'},$$

so the solution of (2.22) exists. $\qquad \square$

Proof of Theorem 2.1 By Theorem 2.2 and noticing that as both $a(x)$ and f are functions with axi-symmetry, according to the uniqueness of solution, if the solution of (2.20) exists, it must be the unique solution of (2.20), which satisfies (2.21).

We now prove the existence of solution of (2.20).

For any given $f \in (L_*^p(\Omega))^2$, by means of mollifier, there exists $f_\delta \in (C^\infty(\overline{\Omega}))^2$ such that as $\delta \to 0$, f_δ converges to f strongly in $L_*^p(\Omega)$. For any given $f_\delta \in (C^\infty(\overline{\Omega}))^2$, by Lax-Milgram theorem, there exists a solution $v_\delta \in H_*^1(\Omega)$ such that

$$\iint_\Omega \sigma \left(\frac{\partial v_\delta}{\partial r}\frac{\partial\phi}{\partial r} + \frac{\partial v_\delta}{\partial z}\frac{\partial\phi}{\partial z} \right) rdrdz = \iint_\Omega \left(f_{1\delta}\frac{\partial\phi}{\partial r} + f_{2\delta}\frac{\partial\phi}{\partial z} \right) rdrdz, \quad \forall \phi \in H_*^1(\Omega). \qquad (2.39)$$

Evidently for $1 < p < 2$, v_δ is also a $W_*^{1,p}$ solution of (2.39). Putting this solution into 3-D situation, by Theorem 2.2 we know that v_δ satisfies

$$\|v_\delta\|_{W_*^{1,p}(\Omega)} \leq C\|f_\delta\|_{L_*^p(\Omega)} \leq C\|f\|_{L_*^p(\Omega)}. \qquad (2.40)$$

Then, $\{v_\delta\}$ is uniformly bounded with respect to δ in $W_*^{1,p}(\Omega)$, therefore there exists a subsequence $\{v_{\delta_k}\}$ such that as $k \to +\infty$, $v_{\delta_k} \rightharpoonup v$ weakly in $W_*^{1,p}(\Omega)$.

Taking $\delta = \delta_k$ in (2.39) and putting $k \to +\infty$, we obtain

$$\iint_\Omega \sigma \left(\frac{\partial v}{\partial r} \frac{\partial \phi}{\partial r} + \frac{\partial v}{\partial z} \frac{\partial \phi}{\partial z} \right) r \, dr \, dz = \iint_\Omega \left(f_1 \frac{\partial \phi}{\partial r} + f_2 \frac{\partial \phi}{\partial z} \right) r \, dr \, dz, \qquad (2.41)$$

which shows the existence of solution of (2.20).   □

According to Theorem 2.1 and Lemma 2.2, we have

Theorem 2.3 *There exists $\varepsilon_0 > 0$, for any given p $(2 - \varepsilon_0 < p < 2)$, problem (SP) admits a unique solution $u \in V_p$, and*

$$\sum_{s=1}^{4} \|u\|_{W_*^{1,p}(\Omega_s)} \leq C(p) \left(\sum_{k=1}^{5} \|E_k\|_{W^{1-1/p,p}(\gamma_k)} \right). \qquad (2.42)$$

Proof Let $w \in V_p$ satisfy (2.18). By Lemma 2.2, $v = u - w$ should satisfy (2.20). By Theorem 2.1, there exists a unique $v \in V_p^0$ such that (2.20) holds and

$$\|v\|_{W_*^{1,p}(\Omega)} \leq C(p) \|f\|_{L_*^p(\Omega)}, \qquad (2.43)$$

where $f = \sigma_i \nabla w$ in Ω_i $(i = 1, 2, 3, 4)$. The combination of (2.43) and (2.18) gives

$$\sum_{i=1}^{4} \|u\|_{W_*^{1,p}(\Omega_i)} \leq \|v\|_{W_*^{1,p}(\Omega)} + \sum_{i=1}^{4} \|w\|_{W_*^{1,p}(\Omega_i)}$$

$$\leq C(p) \sum_{i=1}^{4} \|w\|_{W_*^{1,p}(\Omega_i)}$$

$$\leq C(p) \sum_{k=1}^{5} \|E_k\|_{W^{1-1/p,p}(\gamma_k)}. \qquad (2.44)$$

It is evident that $u = v + w \in V_p$ is the unique solution of problem (SP).   □

References

1. Adams, R.A.: Sobolev Space. Academic Press, New York (1975)
2. Chen, Y., Wu, L.: Elliptic Partial Differential Equations of Second Order and System of Elliptic Equations. Science Press, Beijing (1991) (in Chinese)
3. Gilbarg, D., Trudinger, N.S.: Elliptic Partial Differential Equations of Second Order. Springer, Berlin (1977)
4. Grisvard, P.: Elliptic Problems in Nonsmooth Domains. Pitman Advanced Publishing Program, Boston (1985)
5. Li, T., Tan, Y., Peng, Y.: Mathematical model and method for the spontaneous potential well-logging. Eur. J. Appl. Math. **5**, 123–139 (1994)

6. Peng, Y.: The necessary and sufficient condition for the well-posedness of a kind PDE. J. Tongji Univ. **16**(1), 91–100 (1988) (in Chinese)
7. Zhou, Y., Cai, Z.: Convergence of a numerical method in mathematical spontaneous potential well-logging. Eur. J. Appl. Math. **7**(1), 31–41 (1996)

Chapter 3
Limiting Behavior

Problem (SP) can be solved numerically by the finite element method (see Chap. 4). Since u satisfies the equi-potential surface conditions (1.9)–(1.10) on the surface of the electrode Γ_0, we must make a special treatment for the equi-potential surface Γ_0 in the computation scheme of finite element method. Noticing that in some practical problems, if the measure of the equi-potential surface Γ_0 is relatively small and the measuring electrode does not emit any electric current, we can regard the equi-potential surface Γ_0 as a point and solve only a usual elliptic boundary value problem (SP) without equi-potential boundary condition in the whole domain $\{(r, z) \mid 0 \leq r \leq R, 0 \leq z \leq Z\}$ (see Fig. 3.1). In this way the mesh partition on the domain is much simpler, the special treatment for the equi-potential boundary can be avoided, then it may benefit the computation. However, whether this approximate treatment is reasonable and reliable or not? This problem depends on the research on the limiting behavior of the solution as the domain is varying. We should mention that although in [4] the limiting behavior of the solution has been investigated as a star-shaped equi-potential surface shrinks to a point, but the simple constant extension method used for that case to extend the H^1 function in a domain with hole to the entire domain does not work for our problem (SP). Because the Neumann boundary condition should be satisfied on Γ_{27} (see Fig. 1.1) and the constant extension can not guarantee that the extended function belongs to H^1 in the whole domain. Besides, we know that for a general domain $\Omega \subseteq \widetilde{\Omega}$, there always exists an extension operator P such that $Pu \in H^1(\widetilde{\Omega})$, $Pu|_\Omega = u$ and $\|Pu\|_{H^1(\widetilde{\Omega})} \leq C\|u\|_{H^1(\Omega)}$ for any given $u \in H^1(\Omega)$. However, general speaking, the dominant constant C depends on the domain Ω. We will use the particularity of the domain in problem (SP) and find a suitable extension method such that the above dominant constant C no longer depends on the domain. As a result, we can get the limiting behavior of the solution of problem (SP) as the equi-potential surface shrinks to a point, which gives a positive answer to the above question [2].

Lemma 3.1 *As shown in Fig. 3.2, let $\Omega = \Omega_1 \cup \Omega_2$, Γ be the boundary of Ω except the bottom line, γ the interline between Ω_1 and Ω_2, and γ_ε a segment on γ starting*

T. Li et al., *Mathematical Model of Spontaneous Potential Well-Logging and Its Numerical Solutions*, SpringerBriefs in Mathematics, DOI 10.1007/978-3-642-41425-1_3, © The Author(s) 2014

Fig. 3.1 Simplifies model of
the formation

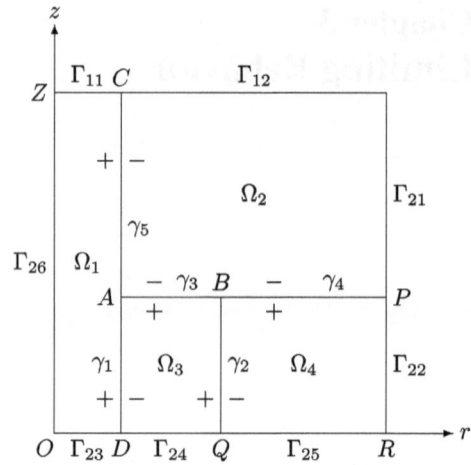

Fig. 3.2 Figure for
Lemma 3.1

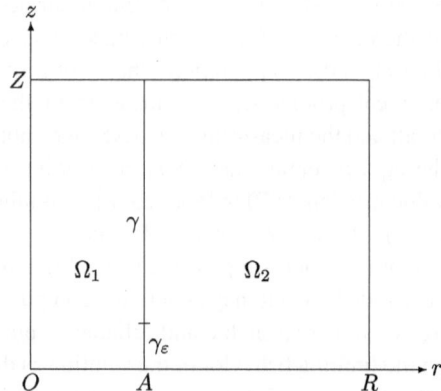

from A with length ε. We denote

$$V = \{v \mid v \in H_*^1(\Omega), v|_\Gamma = f\},\tag{3.1}$$

$$V_\varepsilon = \{v \mid v \in V, v|_{\gamma_\varepsilon} = \text{constant}\},\tag{3.2}$$

then for any given $v \in V$, there exists $v_\varepsilon \in V_\varepsilon$, such that $v_\varepsilon \to v$ strongly in $H_*^1(\Omega)$ as $\varepsilon \to 0$.

Proof We only need to prove that for any given $\eta > 0$, there exists $\varepsilon_0 > 0$, such that for any given $\varepsilon \in (0, \varepsilon_0)$, there exists $v_\varepsilon \in V_\varepsilon$ satisfying $\|v_\varepsilon - v\|_{H_*^1(\Omega)} \leq \eta$.

We denote $B_1 = B_{d/3}(A)$, $B_2 = B_{2d/3}(A)$, where $B_r(A)$ stands for a circle with radius r centered at A, and $d > 0$ is sufficient small. Taking $\phi \in C^\infty(\Omega)$ such that

$$\phi = \begin{cases} 1 & \text{in } \Omega \backslash B_2, \\ 0 & \text{in } B_1 \cap \Omega, \end{cases}\tag{3.3}$$

it is evident that $\phi v \in H^1_*(\Omega)$. Let

$$w \stackrel{\text{def.}}{=} v - \phi v = \begin{cases} 0 & \text{in } \Omega \backslash B_2, \\ v & \text{in } B_1 \cap \Omega. \end{cases} \tag{3.4}$$

It is easy to see that $w \in H^1_*(\Omega)$. By density, there exists $w_\eta \in C^\infty(\overline{\Omega})$ such that $w_\eta|_\Gamma = 0$ and

$$\|w_\eta - w\|_{H^1_*(\Omega)} \le \frac{\eta}{2}. \tag{3.5}$$

Denoting $w_\eta|_{\partial\Omega_1} = f_{1\eta}$, $w_\eta|_{\partial\Omega_2} = f_{2\eta}$, it is evident that $f_{1\eta}|_\gamma = f_{2\eta}|_\gamma$ $(\stackrel{\text{def.}}{=} f_\eta)$. Let

$$f_{\eta\varepsilon}(s) = \begin{cases} f_\eta(s) & \text{on } \gamma \backslash \gamma_{2\varepsilon}, \\ f_\eta(0) + (f_\eta(2\varepsilon) - f_\eta(0))h(\frac{s}{\varepsilon}) & \text{on } \gamma_{2\varepsilon}, \end{cases} \tag{3.6}$$

where s is the arc length parameter which starts from A and orients the positive direction of the z-axis, and

$$h(s) = \begin{cases} 0 & s \le 1 \\ 1 & s \ge 2 \end{cases} \in C^\infty(R^+). \tag{3.7}$$

It is evident that $f_{\eta\varepsilon}(s)$ is continuous, and it takes a constant value $f_\eta(0)$ on γ_ε. It is easy to see that $f'_{\eta\varepsilon}(s)$ is uniformly bounded with respect to ε on $\gamma_{2\varepsilon}$, then, by (3.6), $f_{\eta\varepsilon}(s)$ is a Lipschitz continuous function of s on γ uniformly with respect to the parameter ε.

Denoting

$$g_{\eta\varepsilon} = f_{\eta\varepsilon} - f_\eta, \tag{3.8}$$

we have [1]

$$\|g_{\eta\varepsilon}\|^2_{H^{1/2}(\gamma)}$$

$$= \iint_{\gamma \times \gamma} \left| \frac{g_{\eta\varepsilon}(x) - g_{\eta\varepsilon}(y)}{x - y} \right|^2 dx dy$$

$$= \left(\iint_{(\gamma \backslash \gamma_{2\varepsilon}) \times (\gamma \backslash \gamma_{2\varepsilon})} + \iint_{(\gamma \backslash \gamma_{2\varepsilon}) \times \gamma_{2\varepsilon}} + \int\int_{\gamma_{2\varepsilon} \times (\gamma \backslash \gamma_{2\varepsilon})} + \iint_{\gamma_{2\varepsilon} \times \gamma_{2\varepsilon}} \right)$$

$$\left| \frac{g_{\eta\varepsilon}(x) - g_{\eta\varepsilon}(y)}{x - y} \right|^2 dx dy$$

$$\stackrel{\text{def.}}{=} I_1 + I_2 + I_3 + I_4.$$

Since $g_{\eta\varepsilon} \equiv 0$ on $\gamma \backslash \gamma_{2\varepsilon}$, $I_1 = 0$, and $g_{\eta\varepsilon}$ is a Lipschitz continuous function of s on γ uniformly with respect to the parameter ε, obviously we have $I_2, I_3, I_4 \le C\varepsilon$.

Therefore, we obtain

$$\|g_{\eta\varepsilon}\|_{H^{1/2}(\gamma)} \le C\sqrt{\varepsilon},$$

i.e.,

$$\|f_{\eta\varepsilon} - f_{\eta}\|_{H^{1/2}(\gamma)} \le C\sqrt{\varepsilon}. \tag{3.9}$$

By the inverse trace theorem [3], and noticing (3.9), there exist $u_{1\varepsilon} \in H_*^1(\Omega_1)$ and $u_{2\varepsilon} \in H_*^1(\Omega_2)$ such that

$$2u_{1\varepsilon}|_{\partial\Omega_1\setminus\gamma} = f_{1\eta}|_{\partial\Omega_1\setminus\gamma}, \qquad u_{1\varepsilon}|_{\gamma} = f_{\eta\varepsilon}, \tag{3.10}$$

$$u_{2\varepsilon}|_{\partial\Omega_2\setminus\gamma} = f_{2\eta}|_{\partial\Omega_2\setminus\gamma}, \qquad u_{2\varepsilon}|_{\gamma} = f_{\eta\varepsilon} \tag{3.11}$$

and

$$\|u_{1\varepsilon} - w_{\eta}\|_{H_*^1(\Omega_1)} \le C\sqrt{\varepsilon}, \qquad \|u_{2\varepsilon} - w_{\eta}\|_{H_*^1(\Omega_2)} \le C\sqrt{\varepsilon}. \tag{3.12}$$

Let

$$u_{\varepsilon} = \begin{cases} u_{1\varepsilon} & \text{in } \Omega_1, \\ u_{2\varepsilon} & \text{in } \Omega_2. \end{cases} \tag{3.13}$$

Obviously, $u_{\varepsilon} \in H_*^1(\Omega)$, and by (3.12), $\|u_{\varepsilon} - w_{\eta}\|_{H_*^1(\Omega)} \le C_0\sqrt{\varepsilon}$, where C_0 is a positive constant independent of ε. Taking $\varepsilon_0 = \frac{\eta^2}{4C_0^2}$, for any given $\varepsilon \in (0, \varepsilon_0)$ we have

$$\|u_{\varepsilon} - w_{\eta}\|_{H_*^1(\Omega)} \le \frac{\eta}{2}. \tag{3.14}$$

Let $v_{\varepsilon} = \phi v + u_{\varepsilon}$. It is evident that $v_{\varepsilon} \in V_{\varepsilon}$, and by (3.4)–(3.5) and (3.14), it is easy to get $\|v_{\varepsilon} - v\|_{H_*^1(\Omega)} \le \eta$. $\qquad\square$

Corollary 3.1 *As shown in Fig. 3.3, let $E_k(s)$ ($k = 1, \ldots, 5$) satisfy the compatible condition (2.6), and γ_{ε} be a vertical segment FG with one end point F and length ε. Denoting*

$$W = \{w \mid w \in H_*^1(\Omega_i) \ (i = 1, 2, 3, 4), \ w|_{\Gamma_1} = 0,$$

$$\left(w^+ - w^-\right)\big|_{\gamma_k} = E_k(s) \ (k = 1, \ldots, 5)\}, \tag{3.15}$$

$$W_{\varepsilon} = \{w \mid w \in W, \ w|_{\gamma_{\varepsilon}} = \text{constant}\}, \tag{3.16}$$

then for any given $w \in W$, there exists $w_{\varepsilon} \in W_{\varepsilon}$ such that as $\varepsilon \to 0$, $w_{\varepsilon} \to w$ strongly in $\widetilde{H}_*^1(\Omega)$, where $\widetilde{H}_*^1(\Omega)$ stands for the weighted Sobolev space piecewise defined on domain $\Omega = \bigcup_{i=1}^4 \Omega_i$ with the norm $\|v\|_{\widetilde{H}_*^1(\Omega)} = (\sum_{i=1}^4 \|v\|_{H_*^1(\Omega_i)}^2)^{1/2}$.

Proof By Lemma 2.1, W is not empty. For any given $w \in W$, we denote $w_i = w|_{\Omega_i}$ ($i = 1, 2, 3, 4$). By Lemma 3.1, there exists $w_{1\varepsilon} \in H_*^1(\Omega_1)$, such that $w_{1\varepsilon}|_{\Gamma_{11}} = 0$,

Fig. 3.3 Figure for
Corollary 3.1

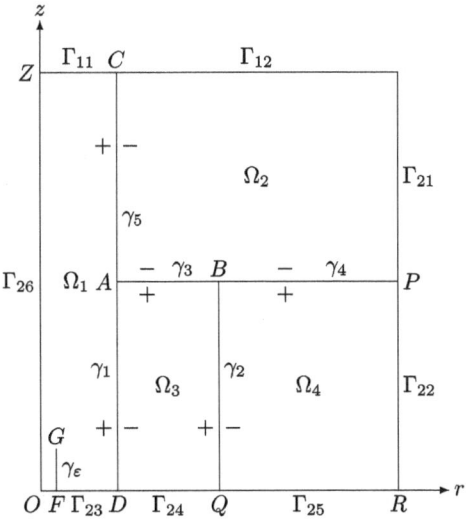

$w_{1\varepsilon}|_{\gamma_k^+} = w_1|_{\gamma_k^+}$ $(k = 1, 5)$, $w_{1\varepsilon}$ is a constant on γ_ε (with length ε), and, as $\varepsilon \to 0$,
$w_{1\varepsilon} \to w_1$ strongly in $H_*^1(\Omega_1)$. Let

$$w_\varepsilon = \begin{cases} w_{1\varepsilon} & \text{in } \Omega_1, \\ w_i & \text{in } \Omega_i \ (i = 2, 3, 4). \end{cases} \tag{3.17}$$

$w_\varepsilon \in W_\varepsilon$ is that we want to find. □

A special case of Corollary 3.1 is the following

Corollary 3.2 *Let*

$$W^0 = \left\{ w \mid w \in H_*^1(\Omega_i) \ (i = 1, 2, 3, 4), \ w|_{\Gamma_1} = 0, \right.$$
$$\left. \left(w^+ - w^- \right)\big|_{\gamma_k} = 0 \ (k = 1, \dots, 5) \right\}$$
$$= \left\{ w \mid w \in H_*^1(\Omega), \ w|_{\Gamma_1} = 0 \right\}, \tag{3.18}$$
$$W_\varepsilon^0 = \left\{ w \mid w \in W^0, \ w|_{\gamma_\varepsilon} = \text{constant} \right\}. \tag{3.19}$$

For any given $w \in W^0$, there exists $w_\varepsilon \in W_\varepsilon^0$, such that as $\varepsilon \to 0$, $w_\varepsilon \to w$ strongly in $H_*^1(\Omega)$.

For operator extension in a general domain, the dominant constant depends on the domain. In what follows we utilize the particularity of the shape of the domain under consideration to find an extension method in which the dominant constant does not depend on the domain.

Fig. 3.4 Figure for
Lemma 3.2

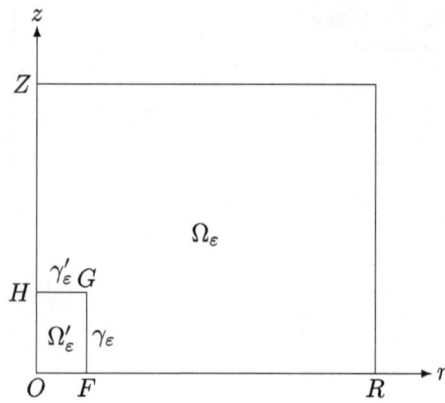

Lemma 3.2 *Let $\Omega = \Omega_\varepsilon \cup \Omega'_\varepsilon$ (see Fig. 3.4). Denoting ε as the length of γ_ε, if $\varepsilon > 0$ is sufficient small, then for any given $u \in H^1_*(\Omega_\varepsilon)$, there exists $\tilde{u} \in H^1_*(\Omega)$ such that $\tilde{u}|_{\Omega_\varepsilon} = u$ and*

$$\|\tilde{u}\|_{H^1_*(\Omega)} \le C \|u\|_{H^1_*(\Omega_\varepsilon)}, \tag{3.20}$$

where C is a positive constant independent of u and ε.

Proof For any given $u \in H^1_*(\Omega_\varepsilon)$, we define

$$\tilde{u}(r,z) = \begin{cases} u(r,z) & \text{in } \Omega_\varepsilon, \\ u(2r_F - r, z) + u(r, 2z_H - z) - u(2r_F - r, 2z_H - z) & \text{in } \Omega'_\varepsilon, \end{cases} \tag{3.21}$$

where r_F is the r-coordinate of point F, z_H is the z-coordinate of point H. It is easy to show that $\tilde{u}$ is continuous on FG and HG, then $\tilde{u} \in H^1_*(\Omega)$, and it is easy to see that

$$\|\tilde{u}\|^2_{H^1_*(\Omega)} = \|\tilde{u}\|^2_{H^1_*(\Omega_\varepsilon)} + \|\tilde{u}\|^2_{H^1_*(\Omega'_\varepsilon)} \le C \|u\|^2_{H^1_*(\Omega_\varepsilon)},$$

so

$$\|\tilde{u}\|_{H^1_*(\Omega)} \le C \|u\|_{H^1_*(\Omega_\varepsilon)},$$

and C is a positive constant independent of both u and ε.   □

Consider the following boundary value problem with equi-potential surface (see Fig. 3.5)

Fig. 3.5 Figure for (3.22)

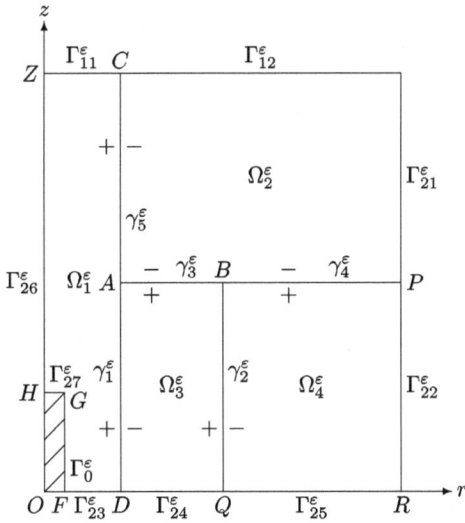

$$\begin{cases} \dfrac{\partial}{\partial r}\left(\sigma r\dfrac{\partial u_\varepsilon}{\partial r}\right) + \dfrac{\partial}{\partial z}\left(\sigma r\dfrac{\partial u_\varepsilon}{\partial z}\right) = h & \text{in } \Omega_i^\varepsilon \ (i=1,2,3,4), \\[2mm] u_\varepsilon = 0 & \text{on } \Gamma_1^\varepsilon, \\[2mm] \sigma r\dfrac{\partial u_\varepsilon}{\partial n} = 0 & \text{on } \Gamma_2^\varepsilon, \\[2mm] u_\varepsilon = \text{a constant to be determined} & \text{on } \Gamma_0^\varepsilon, \\[2mm] \displaystyle\int_{\Gamma_0^\varepsilon} \sigma r\dfrac{\partial u_\varepsilon}{\partial n}\,dS = 0, \\[3mm] u_\varepsilon^+ - u_\varepsilon^- = E_k(s) & \text{on } \gamma_k^\varepsilon \ (k=1,\ldots,5), \\[2mm] \left(\sigma r\dfrac{\partial u_\varepsilon}{\partial n}\right)^+ = \left(\sigma r\dfrac{\partial u_\varepsilon}{\partial n}\right)^- & \text{on } \gamma_k^\varepsilon \ (k=1,\ldots,5) \end{cases} \tag{3.22}$$

and the usual boundary value problem (see Fig. 3.1)

$$\begin{cases} \dfrac{\partial}{\partial r}\left(\sigma r\dfrac{\partial u}{\partial r}\right) + \dfrac{\partial}{\partial z}\left(\sigma r\dfrac{\partial u}{\partial z}\right) = h & \text{in } \Omega_i \ (i=1,2,3,4), \\[2mm] u = 0 & \text{on } \Gamma_1, \\[2mm] \sigma r\dfrac{\partial u}{\partial n} = 0 & \text{on } \Gamma_2, \\[2mm] u^+ - u^- = E_k(s) & \text{on } \gamma_k \ (k=1,\ldots,5), \\[2mm] \left(\sigma r\dfrac{\partial u}{\partial n}\right)^+ = \left(\sigma r\dfrac{\partial u}{\partial n}\right)^- & \text{on } \gamma_k \ (k=1,\ldots,5), \end{cases} \tag{3.23}$$

where $\Omega_i^\varepsilon = \Omega_i$ $(i=2,3,4)$.

Problems (3.22) and (3.23) can be reduced to the equivalent variational problems, respectively. Denoting

$$a_\varepsilon(u, v) = \sum_{i=1}^{4} \iint_{\Omega_i^\varepsilon} \sigma_i r \left(\frac{\partial u}{\partial r} \frac{\partial v}{\partial r} + \frac{\partial u}{\partial z} \frac{\partial v}{\partial z} \right) dr dz, \tag{3.24}$$

$$l_\varepsilon(v) = \sum_{i=1}^{4} \iint_{\Omega_i^\varepsilon} h r v dr dz, \tag{3.25}$$

$$a(u, v) = \sum_{i=1}^{4} \iint_{\Omega_i} \sigma_i r \left(\frac{\partial u}{\partial r} \frac{\partial v}{\partial r} + \frac{\partial u}{\partial z} \frac{\partial v}{\partial z} \right) dr dz, \tag{3.26}$$

$$l(v) = \sum_{i=1}^{4} \iint_{\Omega_i} h r v dr dz \tag{3.27}$$

and $\Omega_\varepsilon = \bigcup_{i=1}^{4} \Omega_i^\varepsilon$, $u_\varepsilon \in W_\varepsilon(\Omega_\varepsilon)$ is the solution of the following variational problem: For any given $\phi_\varepsilon \in W_\varepsilon^0(\Omega_\varepsilon)$ it holds that

$$a_\varepsilon(u_\varepsilon, \phi_\varepsilon) = l_\varepsilon(\phi_\varepsilon), \tag{3.28}$$

where

$$W_\varepsilon(\Omega_\varepsilon) = \left\{ w \mid w \in H_*^1\left(\Omega_i^\varepsilon\right) \ (i = 1, 2, 3, 4), \ w|_{\Gamma_1} = 0, \right.$$
$$\left. \left(w^+ - w^-\right)\big|_{\gamma_k} = E_k(s) \ (k = 1, \ldots, 5), \ w|_{\Gamma_0^\varepsilon} = \text{constant} \right\}, \tag{3.29}$$

$$W_\varepsilon^0(\Omega_\varepsilon) = \left\{ w \mid w \in H_*^1\left(\Omega_i^\varepsilon\right) \ (i = 1, 2, 3, 4), \ w|_{\Gamma_1} = 0, \right.$$
$$\left. \left(w^+ - w^-\right)\big|_{\gamma_k} = 0 \ (k = 1, \ldots, 5), \ w|_{\Gamma_0^\varepsilon} = \text{constant} \right\}$$
$$= \left\{ w \mid w \in H_*^1\left(\Omega^\varepsilon\right), \ w|_{\Gamma_1} = 0, \ w|_{\Gamma_0^\varepsilon} = \text{constant} \right\}. \tag{3.30}$$

Correspondingly, $u \in W$ is the solution of the following variational problem: For any given $\phi \in W^0$ it holds that

$$a(u, \phi) = l(\phi), \tag{3.31}$$

where the definitions of W and W^0 can be found in (3.15) and (3.18).

Theorem 3.1 *Suppose that $E_k(s)$ $(k = 1, \ldots, 5)$ satisfy the compatible condition (2.6) and $E_5(C) = 0$. Suppose furthermore that $l_\varepsilon(v)$ and $l(v)$ are the linear continuous functionals of v in spaces $W_\varepsilon^0(\Omega_\varepsilon)$ and W^0, respectively. Then problems (3.22) and (3.23) admit unique piecewise H_*^1 solutions u_ε and u, respectively, and as $\varepsilon \to 0$ (where ε is the length of Γ_0^ε),*

$$\tilde{u}_\varepsilon \to u \quad \text{strongly in } \tilde{H}_*^1(\Omega), \tag{3.32}$$

where $\tilde{u}_\varepsilon$ is the extension of u_ε in the sense of Lemma 3.2.

Proof For any given $w \in W$, by Corollary 3.1, there exists $w_\varepsilon \in W_\varepsilon$ such that as $\varepsilon \to 0$, $w_\varepsilon \to w$ strongly in $\tilde{H}^1_*(\Omega)$. Denote $v_\varepsilon = u_\varepsilon - w_\varepsilon|_{\Omega_\varepsilon}$. By (3.28), for any given $\phi_\varepsilon \in W^0_\varepsilon(\Omega_\varepsilon)$ we have

$$a_\varepsilon(v_\varepsilon, \phi_\varepsilon) = l_\varepsilon(\phi_\varepsilon) - a_\varepsilon(w_\varepsilon, \phi_\varepsilon). \tag{3.33}$$

Evidently, $v_\varepsilon \in W^0_\varepsilon(\Omega_\varepsilon)$. Taking $\phi_\varepsilon = v_\varepsilon$ in the above formula yields

$$a_\varepsilon(v_\varepsilon, v_\varepsilon) = l_\varepsilon(v_\varepsilon) - a_\varepsilon(w_\varepsilon, v_\varepsilon). \tag{3.34}$$

Then we obtain

$$\|\nabla v_\varepsilon\|^2_{L^2_*(\Omega_\varepsilon)} \leq C\big(\|v_\varepsilon\|_{H^1_*(\Omega_\varepsilon)} + \|\nabla w_\varepsilon\|_{L^2_*(\Omega_\varepsilon)} \|\nabla v_\varepsilon\|_{L^2_*(\Omega_\varepsilon)}\big). \tag{3.35}$$

Since as $\varepsilon \to 0$, w_ε converges to w strongly in $\tilde{H}^1_*(\Omega)$, $\|\nabla w_\varepsilon\|_{L^2_*(\Omega_\varepsilon)}$ is uniformly bounded with respect to ε. By Poincaré inequality and noticing Lemma 3.2, we have

$$\|v_\varepsilon\|_{H^1_*(\Omega_\varepsilon)} \leq \|\tilde{v}_\varepsilon\|_{H^1_*(\Omega)} \leq C\|\nabla \tilde{v}_\varepsilon\|_{L^2_*(\Omega)} \leq C\|\nabla v_\varepsilon\|_{L^2_*(\Omega_\varepsilon)}, \tag{3.36}$$

where $\tilde{v}_\varepsilon$ is the extension of v_ε in the sense of Lemma 3.2. By combining (3.35) and (3.36) we obtain $\|\nabla v_\varepsilon\|_{L^2_*(\Omega_\varepsilon)} \leq C$. Moreover, by Lemma 3.2, $\{\tilde{v}_\varepsilon\}$ is a uniformly bounded sequence in $H^1_*(\Omega)$, then there exists a subsequence $\{\tilde{v}_{\varepsilon_n}\}$ such that $\tilde{v}_{\varepsilon_n} \rightharpoonup v$ weakly in $H^1_*(\Omega)$.

By Corollary 3.2, for any given $\phi \in W^0$, there exists $\phi_\varepsilon \in W^0_\varepsilon$ such that as $\varepsilon \to 0$, $\phi_\varepsilon \to \phi$ strongly in $H^1_*(\Omega)$. Substituting above ϕ_ε into (3.33), for any subsequence $\{\tilde{v}_{\varepsilon_k}\}$ of $\{\tilde{v}_\varepsilon\}$, such that $\tilde{v}_{\varepsilon_k} \rightharpoonup v$ weakly in $H^1_*(\Omega)$, we have

$$a_{\varepsilon_k}(v_{\varepsilon_k}, \phi_{\varepsilon_k}) = l_{\varepsilon_k}(\phi_{\varepsilon_k}) - a_{\varepsilon_k}(w_{\varepsilon_k}, \phi_{\varepsilon_k}). \tag{3.37}$$

Let

$$a_{\varepsilon_k}(v_{\varepsilon_k}, \phi_{\varepsilon_k}) - a(v, \phi) = \left(\iint_{\Omega^\varepsilon_1} \frac{r}{\rho} \nabla v_{\varepsilon_k} \cdot \nabla \phi_{\varepsilon_k} \, dr dz - \iint_{\Omega_1} \frac{r}{\rho} \nabla v \cdot \nabla \phi \, dr dz\right)$$

$$+ \sum_{i=2}^{4} \left(\iint_{\Omega^\varepsilon_i} \frac{r}{\rho} \nabla v_{\varepsilon_k} \cdot \nabla \phi_{\varepsilon_k} \, dr dz - \iint_{\Omega_i} \frac{r}{\rho} \nabla v \cdot \nabla \phi \, dr dz\right)$$

$$= I_1 + I_2 + I_3 + I_4.$$

Noting $\Omega^\varepsilon_i = \Omega_i$ $(i = 2, 3, 4)$, and ϕ_ε converges strongly to ϕ, $\tilde{v}_{\varepsilon_k}$ converges weakly to v in $H^1_*(\Omega^\varepsilon_i) = H^1_*(\Omega_i)$ $(i = 2, 3, 4)$, as $\varepsilon_k \to 0$, I_2, I_3, I_4 converge to 0. On the other hand,

$$I_1 = \iint_{\Omega_1} \frac{r}{\rho}(\nabla \tilde{v}_{\varepsilon_k} \cdot \nabla \phi_{\varepsilon_k} - \nabla v \nabla \phi) \, dr dz - \iint_{\Omega_1 \setminus \Omega^\varepsilon_1} \frac{r}{\rho} \nabla \tilde{v}_{\varepsilon_k} \cdot \nabla \phi_{\varepsilon_k} \, dr dz$$

$$= I_{11} + I_{12}.$$

By the same reason as above, $I_{11} \to 0$ as $\varepsilon_k \to 0$. Furthermore, the L_*^2 norm of $\nabla \tilde{v}_{\varepsilon_k}$ and $\nabla \phi_{\varepsilon_k}$ is uniformly bounded with respect ε, so $I_{12} \to 0$ as $\varepsilon_k \to 0$. Thus we have $a_{\varepsilon_k}(v_{\varepsilon_k}, \phi_{\varepsilon_k}) \to a(v, \phi)$ as $\varepsilon_k \to 0$. In the same way we can prove that $l_{\varepsilon_k}(\phi_{\varepsilon_k}) \to l(\phi)$, $a_{\varepsilon_k}(w_{\varepsilon_k}, \phi_{\varepsilon_k}) \to a(w, \phi)$. Therefore, by (3.37) we obtain

$$a(v, \phi) = l(\phi) - a(w, \phi), \tag{3.38}$$

which means that $u = v + w$ is the solution of (3.23). By the uniqueness of solution, $\tilde{v}_\varepsilon \rightharpoonup v$ weakly in $H_*^1(\Omega)$ for the whole sequence.

Finally, it is easy to see that as $\varepsilon_k \to 0$, we have

$$a_\varepsilon(v_\varepsilon, v_\varepsilon) = l_\varepsilon(v_\varepsilon) - a_\varepsilon(w_\varepsilon, v_\varepsilon) \longrightarrow l(v) - a(w, v) = a(v, v), \tag{3.39}$$

i.e.,

$$\|v_\varepsilon\|_{H_*^1(\Omega_\varepsilon)} \to \|v\|_{H_*^1(\Omega)}. \tag{3.40}$$

Then by the proof of Lemma 3.2, it is easy to show that as $\varepsilon_k \to 0$, $\tilde{v}_\varepsilon \to v$ strongly in $H_*^1(\Omega)$, hence $\tilde{u}_\varepsilon = w_\varepsilon + \tilde{v}_\varepsilon \to w + v = u$ strongly in $\tilde{H}_*^1(\Omega)$. $\qquad\square$

In practical applications, E_k ($k = 1, \ldots, 5$) usually do not satisfy the compatible condition (2.6). By Theorem 2.3, for any given p ($2 - \varepsilon_0 < p < 2$), problems (3.22) and (3.23) admit unique piecewise $W_*^{1,p}$ solutions, respectively, and the corresponding limiting behavior as the equi-potential boundary Γ_0 tends to a point will be given by the following theorem.

Theorem 3.2 *As the compatible condition (2.6) is violated, there exists $\varepsilon_0 > 0$ such that for any given p ($2 - \varepsilon_0 < p < 2$), as the length ε of Γ_0^ε goes to 0,*

$$\tilde{u}_\varepsilon \to u \quad \text{strongly in } \tilde{W}_*^{1,p}(\Omega), \tag{3.41}$$

where $\tilde{W}_*^{1,p}(\Omega)$ stands for the corresponding weighted Sobolev space piecewise defined on the domain $\Omega = \bigcup_{i=1}^4 \Omega_i$.

Proof For the time being we suppose that $E_5(s) \equiv E_3(s) \equiv E_2(s) \equiv 0$, $E_1(s) \equiv \Delta_A$, $E_4(s) \equiv \Delta_B$. Let

$$v_A = \begin{cases} (a_1\theta_A + b_1)f(\rho_A) & 0 \le \theta_A \le \pi/2, \\ (a_2\theta_A + b_2)f(\rho_A) & \pi/2 \le \theta_A \le 3\pi/2, \\ (a_3\theta_A + b_3)f(\rho_A) & 3\pi/2 \le \theta_A \le 2\pi, \end{cases} \tag{3.42}$$

where $\rho_A = \sqrt{(r - r_A)^2 + (z - z_A)^2}$, $\theta_A = \arg((r - r_A) + i(z - z_A))$,

$$f(\rho_A) = \begin{cases} 1 & 0 \le \rho_A \le a_0, \\ 0 & 2a_0 \le \rho_A, \end{cases} \quad a_0 \ll 1, \tag{3.43}$$

$$\frac{a_1}{\rho_2} = \frac{a_2}{\rho_1} = \frac{a_3}{\rho_3} = \frac{2\Delta_A}{\pi(2\rho_1 + \rho_2 + \rho_3)} \overset{\text{def.}}{=} \lambda, \tag{3.44}$$

$$b_1 = 0, \qquad b_2 = \frac{\pi}{2}\lambda(\rho_2 - \rho_1), \qquad b_3 = -2\pi\lambda\rho_3. \tag{3.45}$$

Similarly, we take

$$v_B = \begin{cases} (c_1\theta_B + d_1)g(\rho_B) & 0 \le \theta_B \le \pi, \\ (c_2\theta_B + d_2)g(\rho_B) & \pi \le \theta_B \le 3\pi/2, \\ (c_3\theta_B + d_3)g(\rho_B) & 3\pi/2 \le \theta_B \le 2\pi. \end{cases} \tag{3.46}$$

Let $v_\varepsilon = u_\varepsilon - v_A - v_B$. v_ε satisfies the compatible condition (2.6) (this method is called the removing singularity method [5], see Sect. 4.2.2). By Theorem 3.1, $\tilde{v}_\varepsilon \to v$ strongly in $\tilde{H}^1_*(\Omega)$. It is easy to show that for any fixed p $(2 - \varepsilon_0 < p < 2)$, $v_A, v_B \in \tilde{W}^{1,p}_*$, so $\tilde{u}_\varepsilon = \tilde{v}_\varepsilon + v_A + v_B \to v + v_A + v_B = u$ strongly in $\tilde{W}^{1,p}_*(\Omega)$.

Generally speaking, suppose that w_ε and w are the solutions of (3.22) and (3.23), respectively, which satisfy

$$w_\varepsilon^+ - w_\varepsilon^- = w^+ - w^- = \tilde{E}(s) \overset{\text{def.}}{=} \begin{cases} E_1(s) - \Delta_A & \text{on } \gamma_1, \\ E_2(s) & \text{on } \gamma_2, \\ E_3(s) & \text{on } \gamma_3, \\ E_4(s) - \Delta_B & \text{on } \gamma_4, \\ E_5(s) & \text{on } \gamma_5 \end{cases} \tag{3.47}$$

on the interfaces, hence $\tilde{E}(s)$ satisfies the compatible condition (2.6), and then by Theorem 3.1, $\tilde{w}_\varepsilon \to w$ strongly in $\tilde{H}^1_*(\Omega)$. Let $v_\varepsilon = u_\varepsilon - w_\varepsilon$. Due to the proof given in above half section, for any fixed p $(2 - \varepsilon_0 < p < 2)$, as $\varepsilon \to 0$, $\tilde{v}_\varepsilon \to v$ Strongly in $\tilde{W}^{1,p}_*(\Omega)$, then $\tilde{u}_\varepsilon = \tilde{v}_\varepsilon + \tilde{w}_\varepsilon \to v + w = u$ strongly in $\tilde{W}^{1,p}_*(\Omega)$. $\qquad\square$

In what follows we use the numerical simulation to verify the above convergence conclusion. Since the compatibility condition (2.6) is not satisfied in general and the solution possesses certain singularities, we can not directly use the finite element method to solve the problem, instead we need to adopt some methods to remove the singularities. Here we use the transition zone method (see Sect. 4.2.1), which is the most simple and effective method

We use the following elementary geometric parameters: the half thickness of the objective layer $H = 2$ m, the radius of the well-bore $R_0 = 0.125$ m, the radius of the invaded zone $R_{x_0} = 0.65$ m, the radius depth of the domain $R = 1000$ m, the vertical height of the domain $Z = 600$ m.

We choose 4 groups of physical parameters to do the computation, respectively:

(1) $\rho_1 : \rho_2 : \rho_3 : \rho_4 = 1 : 2 : 3 : 4$, $E_1 = 10$ mV, $E_2 = 20$ mV, $E_3 = 100$ mV, $E_4 = 120$ mV, $E_5 = 30$ mV;

(2) $\rho_1 : \rho_2 : \rho_3 : \rho_4 = 1 : 10 : 50 : 100$, $E_1 = 10$ mV, $E_2 = 20$ mV, $E_3 = 100$ mV, $E_4 = 120$ mV, $E_5 = 30$ mV;

Table 3.1 Numerical results on the limiting behavior

ε	First group $U^{(1)} = 149.2300$		Second group $U^{(2)} = 129.3700$		Third group $U^{(3)} = 208.7700$		Fourth group $U^{(4)} = 175.9162$	
	$U_\varepsilon^{(1)}$	RE	$U_\varepsilon^{(2)}$	RE	$U_\varepsilon^{(3)}$	RE	$U_\varepsilon^{(4)}$	RE
0.05	148.8204	0.27 %	126.9033	1.91 %	208.1057	0.32 %	171.7691	2.36 %
0.10	148.8089	0.28 %	126.8515	1.95 %	208.0874	0.33 %	171.6831	2.41 %
0.15	148.7934	0.29 %	126.7816	2.00 %	208.0629	0.34 %	171.5675	2.47 %
0.20	148.7740	0.31 %	126.6936	2.07 %	208.0320	0.35 %	171.4220	2.55 %
0.25	148.7506	0.32 %	126.5869	2.15 %	207.9947	0.37 %	171.2460	2.65 %

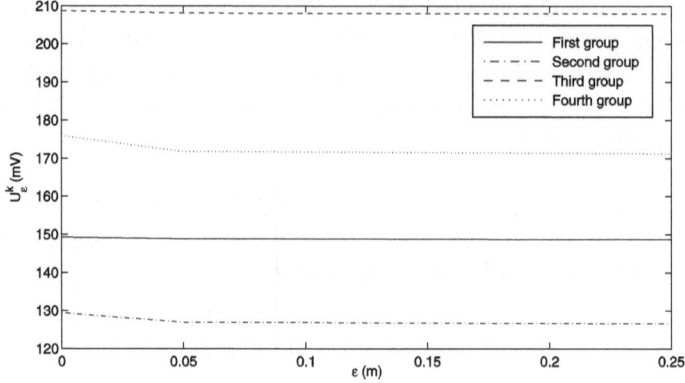

Fig. 3.6 Figure of numerical results on the limiting behavior

(3) $\rho_1 : \rho_2 : \rho_3 : \rho_4 = 1 : 2 : 3 : 4$, $E_1 = 20$ mV, $E_2 = 50$ mV, $E_3 = 130$ mV, $E_4 = 140$ mV, $E_5 = 10$ mV;

(4) $\rho_1 : \rho_2 : \rho_3 : \rho_4 = 1 : 10 : 50 : 100$, $E_1 = 20$ mV, $E_2 = 50$ mV, $E_3 = 130$ mV, $E_4 = 140$ mV, $E_5 = 10$ mV.

Let $U_\varepsilon^{(k)}$ be the value of spontaneous potential on the measuring electrode with different length(denoted by ε) (where $k = 1, 2, 3, 4$ correspond to above 4 groups of parameters), and $U^{(k)}$ the value of spontaneous potential on the measuring electrode as $\varepsilon = 0$ (i.e., at the origin of the well axis). The numerical results are listed in Table 3.1, where the relative error (RE) means the relative error between $U_\varepsilon^{(k)}$ and $U^{(k)}$:

$$\left| \frac{U_\varepsilon^{(k)} - U^{(k)}}{U^{(k)}} \right| \times 100 \%.$$

The results show that the conclusion of above limiting behavior fits engineering applications.

Figure 3.6 shows the picture of the value $U_\varepsilon^{(k)}$ ($k = 1, 2, 3, 4$) varying along with the electrode length, from which we can see that when the length of the electrode decreases, the solution is convergent. Therefore in the numerical simulation, for those electrodes with smaller length, we can shrink the electrode to a limiting point (i.e., the origin of the well axis) to reduce the cost of meshing.

References

1. Adams, R.A.: Sobolev Space. Academic Press, New York (1975)
2. Cai, Z.: Asymptotic behavior for a class of elliptic equivalued surface boundary value problem with discontinuous interface conditions. Appl. Math. J. Chin. Univ. Ser. B **10**(3), 237–250 (1995)
3. Grisvard, P.: Elliptic Problems in Nonsmooth Domains. Pitman Advanced Publishing Program, Boston (1985)
4. Li, T., Chen, S.: Limiting behavior of the equi-valued surface boundary value problem for second order selfadjoint elliptic equations. J. Fudan Univ. **4**, 6–14 (1978) (in Chinese)
5. Li, T., Tan, Y., Peng, Y.: Mathematical model and method for the spontaneous potential well-logging. Eur. J. Appl. Math. **5**, 123–139 (1994)

Chapter 4
Techniques of Solution

4.1 Finite Element Method (FEM) in the Case of Compatible Condition (2.6) Being Valid

We first consider the technique to solve the variational problem (2.4) in the case of compatible condition (2.6) being valid. By Sect. 2.2, in the case the variational problem (2.4) admits a unique piecewise H_*^1 solution $u \in V_2$.

Without loss of generality, suppose that we do the triangular partition on the domain Ω (the following method works for the partition of meshes with other shapes), then the variational form (2.4) can be rewritten to

$$\sum_e \iint_e \sigma_e \nabla u \cdot \nabla \phi r \, dr \, dz = 0, \tag{4.1}$$

where e stands for any triangular element, and $\sigma_e = \sigma_i$, if $e \subseteq \Omega_i$.

For any given triangular element e, we denote the coordinates of its three vertices in the counterclockwise order by (r_p, z_p) $(p = 1, 2, 3)$, and the values of function u at the vertices by u_p $(p = 1, 2, 3)$. We do the linear interpolation for function u:

$$u = \sum_{p=1}^{3} N_p(r, z) u_p, \tag{4.2}$$

where

$$N_p(r, z) = a_p r + b_p z + c_p \quad (p = 1, 2, 3) \tag{4.3}$$

are shape functions with coefficients a_p, b_p and c_p $(p = 1, 2, 3)$ given by

$$\begin{cases} a_1 = \dfrac{1}{2\Delta}(z_2 - z_3), & b_1 = \dfrac{1}{2\Delta}(r_3 - r_2), & c_1 = \dfrac{1}{2\Delta}(r_2 z_3 - r_3 z_2), \\[2mm] a_2 = \dfrac{1}{2\Delta}(z_3 - z_1), & b_2 = \dfrac{1}{2\Delta}(r_1 - r_3), & c_2 = \dfrac{1}{2\Delta}(r_3 z_1 - r_1 z_3), \\[2mm] a_3 = \dfrac{1}{2\Delta}(z_1 - z_2), & b_3 = \dfrac{1}{2\Delta}(r_2 - r_1), & c_3 = \dfrac{1}{2\Delta}(r_1 z_2 - r_2 z_1), \end{cases} \tag{4.4}$$

T. Li et al., *Mathematical Model of Spontaneous Potential Well-Logging and Its Numerical Solutions*, SpringerBriefs in Mathematics, DOI 10.1007/978-3-642-41425-1_4, © The Author(s) 2014

and

$$\Delta = \frac{1}{2}\begin{vmatrix} 1 & r_1 & z_1 \\ 1 & r_2 & z_2 \\ 1 & r_3 & z_3 \end{vmatrix} = \frac{1}{2}(r_1 z_2 + r_2 z_3 + r_3 z_1 - r_1 z_3 - r_2 z_1 - r_3 z_2) \qquad (4.5)$$

is the area of element e. From it we obtain the element stiffness matrix:

$$A^e = \left(\iint_e \sigma_e \nabla N_p \cdot \nabla N_q r dr dz \right)_{3 \times 3}$$

$$= \sigma_e \frac{\Delta}{3}(r_1 + r_2 + r_3) \begin{pmatrix} a_1^2 + b_1^2 & a_1 a_2 + b_1 b_2 & a_1 a_3 + b_1 b_3 \\ a_2 a_1 + b_2 b_1 & a_2^2 + b_2^2 & a_2 a_3 + b_2 b_3 \\ a_3 a_1 + b_3 b_1 & a_3 a_2 + b_3 b_2 & a_3^2 + b_3^2 \end{pmatrix}. \qquad (4.6)$$

We assemble the element stiffness matrices A^e into the total stiffness matrix A which is a symmetric matrix.

Since there are spontaneous potential differences on the interfaces γ_k ($k = 1, \ldots, 5$), we have to make a corresponding treatment. For fixing the idea, we suppose that only the potential value on γ_k^+ is reserved, while the potential value on γ_k^- will be represented by the potential value on γ_k^+ through the interface condition (1.12).

For any given element e, if the node point $q \in \gamma_k^+$, we take it as a common node point to which no extra treatment is needed since u_q^+ (denoted by u_q) is kept as a freedom. If the node point $q \in \gamma_k^-$, by the interface condition (1.12), we have

$$u_q^- = u_q - E_k(q),$$

the contribution of which to the variational form is then

$$\iint_e \sigma_e \nabla N_p \cdot \nabla N_q \big(u_q - E_k(q) \big) r dr dz$$

$$= \left(\iint_e \sigma_e \nabla N_p \cdot \nabla N_q r dr dz \right) u_q - \left(\iint_e \sigma_e \nabla N_p \cdot \nabla N_q r dr dz \right) E_k(q). \qquad (4.7)$$

The first term of the above formula is the contribution to the element stiffness matrix, and the second term is that to the right-hand side term, which can be written as

$$b^e = A_{\cdot q}^e E_k(q), \qquad (4.8)$$

where $A_{\cdot q}^e$ stands for the q-column of the element stiffness matrix A^e, and $E_k(q)$ stands for the value of the spontaneous potential difference E_k at point q. Assembling b^e into the right-hand side term b, we obtain the system of liner algebraic equations

$$AU = b, \qquad (4.9)$$

where $U = (u_1, \ldots, u_N)^\top$ is the column vector formed by the potential values at all the node points (for the node points on the interface γ, we take the potential values on γ^+), and N is the total number of the node points.

We treat the boundary conditions as follows.

For the Dirichlet boundary condition $u|_{\Gamma_1} = 0$, a treatment method is to cancel corresponding rows and columns in the matrix A and in the right-hand side vector b [4, 5]: For any given node point k on Γ_1, we cancel the kth row and kth column of the matrix A and at the same time we cancel the kth entry of the right-hand side vector b. However, in this way we have to move a great amount of data in computation, and it is rather time consuming.

To avoid such movement of data we may adopt another treatment method: For any given node point k on Γ_1, we take the kth row and the kth column of the coefficient matrix $A = (a_{pq})$ as unit vector ε_k (i.e., the vector with the kth entry 1, and the other entries are all 0), respectively. That is

$$a_{kk} := 1, \qquad a_{kq} := 0 \quad (q \neq k), \qquad a_{pk} := 0 \quad (p \neq k). \tag{4.10}$$

In the meantime, we assign the kth entry of the right-hand side vector b as 0, i.e., $b_k = 0$.

For the equi-potential boundary condition $u|_{\Gamma_0} = $ constant, we treat it as follows: Fixing a node point k on Γ_0, for any other node point p on that equi-potential surface we add the pth row of the matrix A to its kth row and add its pth column to the kth column, then assign the pth row and pth column as unit vector ε_p, i.e.,

$$\begin{aligned} a_{kq} := a_{kq} + a_{pq}, \qquad a_{qk} := a_{qk} + a_{qp} \quad (q = 1, \ldots, N), \\ a_{pp} := 1, \qquad a_{pq} := 0, \qquad a_{qp} := 0 \quad (q \neq p). \end{aligned} \tag{4.11}$$

After this treatment, the symmetry of matrix A is still reserved.

Then, we add the pth entry of the right-hand side vector b to its kth entry, then assign its pth entry to be 0, i.e.,

$$b_k := b_k + b_p, \qquad b_p := 0. \tag{4.12}$$

After having made the treatment on the boundary conditions, we obtain the following system of linear algebraic equations:

$$\tilde{A}U = \tilde{b}, \tag{4.13}$$

where $\tilde{A}$ and $\tilde{b}$ are the total stiffness matrix and the right-hand side vector after the treatment. System (4.13) is the FEM discrete scheme of the variational problem (2.4). Solving this system of equations, we will obtain the values of potential function at all the node points, especially the value of spontaneous potential on the electrode A_0.

4.2 Solution Techniques as the Compatible Condition (2.6) Is Violated

If the compatible condition (2.6) is violated, by Lemma 2.1, $V_2 = \emptyset$, then the variational problem (2.4) does not admit a piecewise H_*^1 solution. However, by Theorem 2.3, there exists $\varepsilon_0 > 0$ such that for any fixed p satisfying $2 - \varepsilon_0 < p < 2$, the variational problem (2.4) admits a unique piecewise $W_*^{1,p}$ solution $u \in V_p$.

Since the occurrence of singularity, we can not directly use the FEM to solve (2.4), and it is necessary for us to adopt some techniques to remove singularities. There are three methods: the transition zone method (see Sect. 4.2.1), the removing singularity method (see Sect. 4.2.2) and the double layer potential method (see Sect. 4.2.3).

4.2.1 Transition Zone Method

We introduce the transition zone method (also called the compatibility method of interface conditions, see [6]). It is the simplest and most efficient method.

For $\varepsilon > 0$ sufficient small, we define

$$
\begin{cases}
E_k^\varepsilon(s) = E_k(s) g\left(\dfrac{s}{\varepsilon}\right), & k = 1, 2, 4, 5, \\[2mm]
E_3^\varepsilon(s) = E_3(s) g\left(\dfrac{s}{\varepsilon}\right) g\left(\dfrac{r_B - r_A - s}{\varepsilon}\right),
\end{cases}
\tag{4.14}
$$

where r_A and r_B are the r-coordinates of A and B, respectively,

$$
g(\delta) = \begin{cases}
\delta, & 0 \le \delta \le 1, \\
1, & \delta > 1.
\end{cases}
\tag{4.15}
$$

It is easy to see that we only properly change the initial spontaneous potential differences $E_k(s)$ ($k = 1, \ldots, 5$) in a neighborhood of the crossing points A and B, such that the compatible condition (2.6) is satisfied. Then, by Sect. 2.2, the problem corresponding to $E_k^\varepsilon(s)$ ($k = 1, \ldots, 5$) admits a unique piecewise H_*^1 solution u_ε, and it can be solved by FEM introduced in Sect. 4.1. As ε is small enough, we can regard u_ε as an approximate solution to the original problem (2.4). The rationality of this method can be guaranteed by the following Theorem 4.1 [8].

Theorem 4.1 *There exists $\varepsilon_0 > 0$ such that for any given p ($2 - \varepsilon_0 < p < 2$), the solution u_ε given by the transition zone method converges strongly to the solution u of the original problem in $W_*^{1,p}$ as $\varepsilon \to 0$.*

Proof Since u_ε is the piecewise H_*^1 solution of problem (2.4), where $E_k = E_k^\varepsilon$ ($k = 1, \ldots, 5$) defined by (4.13) satisfy the compatible condition (2.6). By Theorem 2.3, there exists $\varepsilon_0 > 0$, such that for any given p $(2 - \varepsilon_0 < p < 2)$ we have the following L^p estimation for u_ε:

$$\sum_{i=1}^{4} \|u_\varepsilon\|_{W_*^{1,p}(\Omega_i)} \le C(p) \left(\sum_{k=1}^{5} \|E_k^\varepsilon\|_{W^{1-1/p,p}(\gamma_k)} \right). \tag{4.16}$$

Then, by the linear superposition we get

$$\sum_{i=1}^{4} \|u_\varepsilon - u\|_{W_*^{1,p}(\Omega_i)} \le C(p) \left(\sum_{k=1}^{5} \|E_k^\varepsilon - E_k\|_{W^{1-1/p,p}(\gamma_k)} \right). \tag{4.17}$$

So $u_\varepsilon \to u$ strongly in $W_*^{1,p}$ $(2 - \varepsilon_0 < p < 2)$. $\qquad\square$

4.2.2 Removing Singularity Method

In the case that the compatible condition (2.6) is violated, another method of treatment is the removing singularity method [7], The main idea is to construct a function w, such that the jump values of w at point A and point B just satisfy

$$\left(w^+ - w^-\right)\big|_{\gamma_1}(A) + \left(w^+ - w^-\right)\big|_{\gamma_3}(A) - \left(w^+ - w^-\right)\big|_{\gamma_5}(A) = \Delta_A, \tag{4.18}$$

$$\left(w^+ - w^-\right)\big|_{\gamma_2}(B) - \left(w^+ - w^-\right)\big|_{\gamma_3}(B) + \left(w^+ - w^-\right)\big|_{\gamma_4}(B) = \Delta_B. \tag{4.19}$$

Then, denoting $v = u - w$, we rewrite the variational problem (2.4) into the corresponding form with respect to v, while v satisfies the compatible condition (2.6).

The method to construct w can be introduced as follows.

Let

$$w_A = \begin{cases} (a_1\theta_A + b_1)f(\rho_A) & 0 \le \theta_A \le \pi/2, \\ (a_2\theta_A + b_2)f(\rho_A) & \pi/2 \le \theta_A \le 3\pi/2, \\ (a_3\theta_A + b_3)f(\rho_A) & 3\pi/2 \le \theta_A \le 2\pi, \end{cases} \tag{4.20}$$

where

$$\rho_A = \sqrt{(r - r_A)^2 + (z - z_A)^2}, \tag{4.21}$$

$$\theta_A = \arg\big((r - r_A) + \mathrm{i}(z - z_A)\big), \tag{4.22}$$

$$f(\rho_A) = \begin{cases} 1 & 0 \le \rho_A \le a_0, \\ 0 & 2a_0 \le \rho_A, \end{cases} \quad a_0 \ll 1, \tag{4.23}$$

$$\frac{a_1}{\rho_2} = \frac{a_2}{\rho_1} = \frac{a_3}{\rho_3} = \frac{2\Delta_A}{\pi(2\rho_1 + \rho_2 + \rho_3)} \overset{\text{def.}}{=} \lambda, \tag{4.24}$$

$$b_1 = 0, \qquad b_2 = \frac{\pi}{2}\lambda(\rho_2 - \rho_1), \qquad b_3 = -2\pi\lambda\rho_3. \tag{4.25}$$

It is easy to verify that

$$\left(w^+ - w^-\right)\big|_{\gamma_1}(A) = \Delta_A, \quad \left(w^+ - w^-\right)\big|_{\gamma_3}(A) = 0, \quad \left(w^+ - w^-\right)\big|_{\gamma_5}(A) = 0,$$
(4.26)

then (4.18) is satisfied.

Similarly, choosing properly

$$w_B = \begin{cases} (c_1\theta_B + d_1)g(\rho_B) & 0 \le \theta_B \le \pi, \\ (c_2\theta_B + d_2)g(\rho_B) & \pi \le \theta_B \le 3\pi/2, \\ (c_3\theta_B + d_3)g(\rho_B) & 3\pi/2 \le \theta_B \le 2\pi, \end{cases}$$
(4.27)

w_B satisfies (4.19).

Let $v = u - w_A - w_B$. We transform the variational problem (2.4) into the corresponding form of v. Since v satisfies the compatible condition (2.6), we can solve it by the FEM introduced in Sect. 4.1. Finally, by setting $u = v + w_A + w_B$ we obtain the solution of the original variational problem (2.4).

4.2.3 Double Layer Potential Method

The third method to treat singularity is the double layer potential method [1, 9].

In practical applications the spontaneous potential differences E_k $(k = 1, \ldots, 5)$ are often supposed to be constants. By the description before, using transformation (1.10), we can transform the interface conditions into

$$\left(\tilde{u}^+ - \tilde{u}^-\right)\big|_{\gamma_k} = F_k \quad (k = 1, \ldots, 5),$$
(4.28)

where

$$F_1 = F_2 = F_5 = 0, \qquad F_3 = E_1 + E_3 - E_5 = \Delta_A,$$
$$F_4 = E_1 + E_2 + E_4 - E_5 = \Delta_A + \Delta_B.$$
(4.29)

In this way, $\tilde{u}$ is continuous on the vertical interfaces γ_1, γ_2 and γ_5. For the convenience of statement, we still use u to stand for $\tilde{u}$.

We go back to the 3-D formation model to consider the problem. If the medium is uniform in the whole formation, by the definition of double layer potential [2, 3], the potential at any given point M of the space, produced by the dipole double layer plane Σ with surface density of unit dipole moment is

$$u(M) = \int_\Sigma \frac{\partial}{\partial n}\left(\frac{1}{r_{PM}}\right)dS_P,$$
(4.30)

where r_{PM} is the distance from M to some point P on Σ, dS_P is the infinitesimal area element on double layer plane Σ, n is the unit normal vector which points from the plane Σ^- occupied by the negative electric charges to the plane Σ^+ occupied by the positive electric charges (see Fig. 4.1).

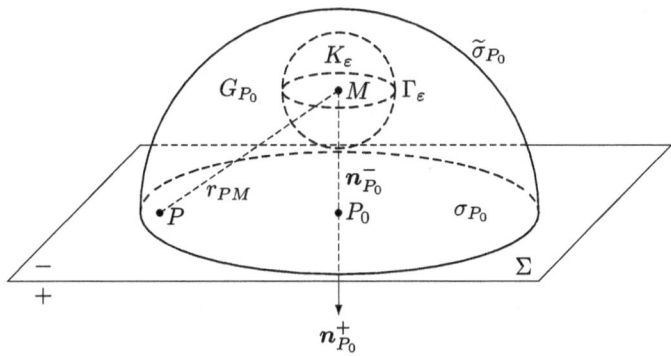

Fig. 4.1 Figure for Lemmas 4.2 and 4.3

Double layer potential (4.30) possesses the following properties [2, 3]:

Lemma 4.1 *For any given $M \notin \Sigma$, double layer potential (4.30) is harmonic at point M, i.e.,*

$$\Delta u(M) = 0. \tag{4.31}$$

Lemma 4.2 *Double layer potential (4.30) has the discontinuity of the first kind on Σ, and for any given $P_0 \in \Sigma$, we have*

$$u^+(P_0) - u^-(P_0) = 4\pi. \tag{4.32}$$

Lemma 4.3 *For any given $P_0 \in \Sigma$, the outer normal derivative of the double layer potential (4.30) is continuous at point P_0, i.e.,*

$$\left(\frac{\partial u}{\partial n}\right)^+ (P_0) = \left(\frac{\partial u}{\partial n}\right)^- (P_0). \tag{4.33}$$

Proof of Lemmas 4.2 and 4.3 We make a suitable large sphere centered at P_0, denote the cross section of it with Σ by σ_{P_0}, the semi-sphere located on the side of surface Σ^- by $\tilde{\sigma}_{P_0}$, and the domain enclosed by them by G_{P_0} (see Fig. 4.1).

As $M \in G_{P_0} \cap n_{P_0}^-$, we make a sphere K_ε ($\subseteq G_{P_0}$) centered at M with sufficiently small radius ε and denote it by Γ_ε (see Fig. 4.1). It is evident that $\frac{1}{r_{PM}}$ is harmonic everywhere in $P \in G_{P_0} \backslash K_\varepsilon$, By the property of harmonic function, we have

$$\int_{\sigma_{P_0} \cup \tilde{\sigma}_{P_0} \cup \Gamma_\varepsilon} \frac{\partial}{\partial n}\left(\frac{1}{r_{PM}}\right) dS_P = 0, \tag{4.34}$$

where n is the outer normal direction of the shown domain. Therefore we have

$$\int_{\sigma_{P_0}} \frac{\partial}{\partial n}\left(\frac{1}{r_{PM}}\right) dS_P = -\int_{\tilde{\sigma}_{P_0}} \frac{\partial}{\partial n}\left(\frac{1}{r_{PM}}\right) dS_P - \int_{\Gamma_\varepsilon} \frac{\partial}{\partial n}\left(\frac{1}{r_{PM}}\right) dS_P. \tag{4.35}$$

Moreover, we have

$$\int_{\Gamma_\varepsilon} \frac{\partial}{\partial n}\left(\frac{1}{r_{PM}}\right) dS_P = -\int_{\Gamma_\varepsilon} \frac{\partial}{\partial r}\left(\frac{1}{r_{PM}}\right) dS_P = \int_{\Gamma_\varepsilon} \frac{1}{r_{PM}^2} dS_P = \int_{\Gamma_\varepsilon} \frac{1}{\varepsilon^2} dS_P = 4\pi.$$

$$(4.36)$$

Then

$$\int_{\sigma_{P_0}} \frac{\partial}{\partial n}\left(\frac{1}{r_{PM}}\right) dS_P = -\int_{\tilde{\sigma}_{P_0}} \frac{\partial}{\partial n}\left(\frac{1}{r_{PM}}\right) dS_P - 4\pi. \qquad (4.37)$$

As $M \in n_{P_0}^+$, $\frac{1}{r_{PM}}$ is harmonic everywhere for $P \in G_{P_0}$, still by the property of harmonic function, we have

$$\int_{\sigma_{P_0}} \frac{\partial}{\partial n}\left(\frac{1}{r_{PM}}\right) dS_P = -\int_{\tilde{\sigma}_{P_0}} \frac{\partial}{\partial n}\left(\frac{1}{r_{PM}}\right) dS_P. \qquad (4.38)$$

As $P \in (\Sigma \backslash \sigma_{P_0}) \cup \tilde{\sigma}_{P_0}$, and $M \in n_{P_0}^\pm$, $\frac{1}{r_{PM}}$ is not of singularity, then $\int_{\Sigma \backslash \sigma_{P_0}} \frac{\partial}{\partial n}(\frac{1}{r_{PM}}) dS_P$ and $\int_{\tilde{\sigma}_{P_0}} \frac{\partial}{\partial n}(\frac{1}{r_{PM}}) dS_P$ is continuous with respect to $M \in n_{P_0}^\pm$, and can be differentiable any times. By (4.30) we have

$$u(M) = \int_\Sigma \frac{\partial}{\partial n}\left(\frac{1}{r_{PM}}\right) dS_P = \int_{\Sigma \backslash \sigma_{P_0}} \frac{\partial}{\partial n}\left(\frac{1}{r_{PM}}\right) dS_P + \int_{\sigma_{P_0}} \frac{\partial}{\partial n}\left(\frac{1}{r_{PM}}\right) dS_P,$$

$$(4.39)$$

then, noticing (4.37) and (4.38), it leads (4.32) then Lemma 4.2.

By (4.37)–(4.39), and noticing the differentiability of

$$\int_{\Sigma \backslash \sigma_{P_0}} \frac{\partial}{\partial n}\left(\frac{1}{r_{PM}}\right) dS_P \quad \text{and} \quad \int_{\tilde{\sigma}_{P_0}} \frac{\partial}{\partial n}\left(\frac{1}{r_{PM}}\right) dS_P,$$

we have

$$\lim_{\substack{M \to P_0 \\ M \in n_{P_0}^+}} \frac{\partial}{\partial n_{P_0}} \int_\Sigma \frac{\partial}{\partial n}\left(\frac{1}{r_{PM}}\right) dS_P = \lim_{\substack{M \to P_0 \\ M \in n_{P_0}^-}} \frac{\partial}{\partial n_{P_0}} \int_\Sigma \frac{\partial}{\partial n}\left(\frac{1}{r_{PM}}\right) dS_P, \qquad (4.40)$$

which is just (4.33), then Lemma 4.3 is true. □

To meet the need of spontaneous potential well-logging, we give the specific expression of double layer potential $u(M)$ below as the surface Σ is a torus perpendicular to the borehole axis. Let $\Sigma = \{(\xi, \eta, \zeta) \mid a_1 \le \sqrt{\xi^2 + \eta^2} \le a_2, \zeta = z_0\}$, where a_1, a_2 $(0 < a_1 < a_2)$ and z_0 are all constants. Suppose that the positive charges distribute on the bottom surface of Σ and the negative charges distribute on the upper surface of Σ (see γ_3 and γ_4 in Fig. 3.1), then by (4.30) at point $M(x, y, z)$ the potential given by the double layer potential is

$$u(x, y, z) = -\iint_{a_1 \le \sqrt{\xi^2+\eta^2} \le a_2} \frac{z - z_0}{((x - \xi)^2 + (y - \eta)^2 + (z - z_0)^2)^{\frac{3}{2}}} d\xi d\eta.$$

By making the polar coordinates transformation

$$\begin{cases} \xi = \rho \cos \varphi, \\ \eta = \rho \sin \varphi, \end{cases}$$

it is easy to see that

$$u(x, y, z) = -\int_0^{2\pi} d\varphi \int_{a_1}^{a_2} \frac{(z - z_0)\rho}{(r^2 + \rho^2 - 2r\rho \cos \varphi + (z - z_0)^2)^{\frac{3}{2}}} d\rho, \qquad (4.41)$$

where $r = \sqrt{x^2 + y^2}$.

Lemma 4.4 *For the double layer potential given by (4.41), as $r = 0$ we have*

$$\frac{\partial u}{\partial r} = 0; \qquad (4.42)$$

as $z = Z$ we have

$$|u| \leq \frac{\pi(a_2^2 - a_1^2)}{(Z - z_0)^2}; \qquad (4.43)$$

and as $r = R$ we have

$$\left| \frac{\partial u}{\partial r} \right| \leq \frac{3\pi(a_2^2 - a_1^2)}{(R - a_2)^3}, \qquad (4.44)$$

where $R > a_2$, $Z > z_0$.

Proof By

$$\frac{\partial u(x, y, z)}{\partial r} = \int_0^{2\pi} d\varphi \int_{a_1}^{a_2} \frac{3(r - \rho \cos \varphi)(z - z_0)\rho}{(r^2 + \rho^2 - 2r\rho \cos \varphi + (z - z_0)^2)^{\frac{5}{2}}} d\rho, \qquad (4.45)$$

we know that

$$\begin{aligned} \frac{\partial u(x, y, z)}{\partial r}\bigg|_{r=0} &= \int_0^{2\pi} d\varphi \int_{a_1}^{a_2} \frac{3(-\rho \cos \varphi)(z - z_0)\rho}{(\rho^2 + (z - z_0)^2)^{\frac{5}{2}}} d\rho \\ &= -3 \int_0^{2\pi} \cos \varphi d\varphi \int_{a_1}^{a_2} \frac{(z - z_0)\rho^2}{(\rho^2 + (z - z_0)^2)^{\frac{5}{2}}} d\rho \\ &= 0, \end{aligned}$$

then (4.42) is valid.

Since $Z > z_0$ and

$$r^2 + \rho^2 - 2r\rho \cos \varphi \geq r^2 + \rho^2 - 2r\rho = (r - \rho)^2 \geq 0, \qquad (4.46)$$

we have

$$\left| u(x, y, Z) \right| = \int_0^{2\pi} d\varphi \int_{a_1}^{a_2} \frac{(Z - z_0)\rho}{(r^2 + \rho^2 - 2r\rho \cos\varphi + (Z - z_0)^2)^{\frac{3}{2}}} d\rho$$

$$\leq \int_0^{2\pi} d\varphi \int_{a_1}^{a_2} \frac{\rho}{(Z - z_0)^2} d\rho$$

$$\leq \frac{\pi(a_2^2 - a_1^2)}{(Z - z_0)^2},$$

i.e., (4.43) is valid.

By (4.45) and $r^2 + \rho^2 - 2r\rho \cos\varphi \geq (r - \rho \cos\varphi)^2$, and noting $R > a_2, a_1 \leq \rho \leq a_2$ and (4.46), we have

$$\left| \frac{\partial u}{\partial r} \right|_{r=R} = \left| \int_0^{2\pi} d\varphi \int_{a_1}^{a_2} \frac{3(R - \rho \cos\varphi)(z - z_0)\rho}{(R^2 + \rho^2 - 2R\rho \cos\varphi + (z - z_0)^2)^{\frac{5}{2}}} d\rho \right|$$

$$\leq \left| \int_0^{2\pi} d\varphi \int_{a_1}^{a_2} \frac{3(z - z_0)\rho}{(R^2 + \rho^2 - 2R\rho \cos\varphi + (z - z_0)^2)^2} d\rho \right|$$

$$\leq \left| \int_0^{2\pi} d\varphi \int_{a_1}^{a_2} \frac{3(z - z_0)\rho}{((R - a_2)^2 + (z - z_0)^2)^2} d\rho \right|$$

$$\leq \left| \int_0^{2\pi} d\varphi \int_{a_1}^{a_2} \frac{3\rho}{((R - a_2)^2 + (z - z_0)^2)^{\frac{3}{2}}} d\rho \right|$$

$$\leq \frac{3\pi(a_2^2 - a_1^2)}{(R - a_2)^3},$$

which is just (4.44).   □

Theorem 4.2 *Suppose that*

$$u(x, y, z) = - \int_0^{2\pi} d\varphi \int_{a_1}^{a_2} \frac{(z - z_0)\rho}{(r^2 + \rho^2 - 2r\rho \cos\varphi + (z - z_0)^2)^{\frac{3}{2}}} d\rho$$

$$+ \int_0^{2\pi} d\varphi \int_{a_1}^{a_2} \frac{(z + z_0)\rho}{(r^2 + \rho^2 - 2r\rho \cos\varphi + (z + z_0)^2)^{\frac{3}{2}}} d\rho, \qquad (4.47)$$

then we have

$$\Delta u = 0 \quad in \ \{(x, y, z) \mid z > 0\} \backslash \{(x, y, z) \mid a_1 < r < a_2, z = z_0\}, \qquad (4.48)$$

$$\begin{cases} u^+ - u^- = 4\pi \\ \left(\frac{\partial u}{\partial n} \right)^+ = \left(\frac{\partial u}{\partial n} \right)^- \end{cases} \quad on \ \{(x, y, z) \mid a_1 < r < a_2, z = z_0\}, \qquad (4.49)$$

$$\frac{\partial u}{\partial n} = 0 \quad \text{on } \big\{(x, y, z) \mid r = 0, z > 0\big\} \cup \big\{(x, y, z) \mid z = 0\big\}, \tag{4.50}$$

and the following estimations hold:

$$|u| \leq \frac{2\pi(a_2^2 - a_1^2)}{(Z - z_0)^2} \quad \text{on } \big\{(x, y, z) \mid z = Z\big\}, \tag{4.51}$$

$$\left|\frac{\partial u}{\partial n}\right| \leq \frac{6\pi(a_2^2 - a_1^2)}{(R - a_2)^3} \quad \text{on } \big\{(x, y, z) \mid r = R, z > 0\big\}, \tag{4.52}$$

where $R > a_2$, $Z > z_0$.

Proof By Lemmas 4.1–4.4, we know that (4.48)–(4.49) and (4.51)–(4.52) are valid. We now prove (4.50). Since

$$
\begin{aligned}
\frac{\partial u(x, y, z)}{\partial z} &= -\int_0^{2\pi} d\varphi \int_{a_1}^{a_2} \frac{(r^2 + \rho^2 - 2r\rho\cos\varphi - 2(z - z_0)^2)\rho}{(r^2 + \rho^2 - 2r\rho\cos\varphi + (z - z_0)^2)^{\frac{5}{2}}} d\rho \\
&\quad + \int_0^{2\pi} d\varphi \int_{a_1}^{a_2} \frac{(r^2 + \rho^2 - 2r\rho\cos\varphi - 2(z + z_0)^2)\rho}{(r^2 + \rho^2 - 2r\rho\cos\varphi + (z + z_0)^2)^{\frac{5}{2}}} d\rho,
\end{aligned}
$$

we have $\frac{\partial u}{\partial z}|_{z=0} = 0$. By (4.42) we know that $\frac{\partial u}{\partial r}|_{r=0} = 0$, it leads to (4.50).   □

Remark 4.1 Here, $u(x, y, z)$ represented by (4.47) is equivalent to the superposition of two double layer potentials produced by two double layer surfaces with surface density of unit dipole moment on $\Sigma_1 = \{(\xi, \eta, \zeta) \mid a_1 \leq \sqrt{\xi^2 + \eta^2} \leq a_2, \zeta = z_0\}$ (the positive charges distribute on the bottom surface of Σ_1) and $\Sigma_2 = \{(\xi, \eta, \zeta) \mid a_1 \leq \sqrt{\xi^2 + \eta^2} \leq a_2, \zeta = -z_0\}$ (the positive charges distribute on the upper surface of Σ_2).

Now we use the properties described before to find the numerical solution of problem (2.4). Here we can simplify our treatment since in practical applications we will meet the case that the differences of spontaneous potentials on the interfaces are constants, and the size of the measuring electrode is very small and then can be considered as a point at the symmetric axis (see Chap. 3). The outline of the computation is as follows:

Step 1 By means of the properties of the double layer potential, find function w, which satisfies (1.2) and the jump conditions $w^+ - w^- = F_k$ $(k = 1, \ldots, 5)$ on the interfaces, where $F_1 = F_2 = F_5 = 0$ and F_3, F_4 are determined by (4.29).

Step 2 Make the translation $v = \tilde{u} - w$, v still satisfies the original equation (1.2) and is continuous on the interfaces, then we are able to use the corresponding variational problem to solve the problem by FEM.

Step 3 Let $u = u_0 + w + v$. We obtain the solution to the original problem (2.4).

According to the above outline, we will give the details of the algorithm and the corresponding formulas below.

Step 1 Let

$$w_{11}(r,z) = -\frac{1}{4\pi}\int_0^{2\pi} d\varphi \int_{R_0}^{R_{x_0}} \frac{(z-H)\rho}{(r^2+\rho^2-2r\rho\cos\varphi+(z-H)^2)^{\frac{3}{2}}}\,d\rho,$$

$$w_{12}(r,z) = \frac{1}{4\pi}\int_0^{2\pi} d\varphi \int_{R_0}^{R_{x_0}} \frac{(z+H)\rho}{(r^2+\rho^2-2r\rho\cos\varphi+(z+H)^2)^{\frac{3}{2}}}\,d\rho,$$

$$w_{21}(r,z) = -\frac{1}{4\pi}\int_0^{2\pi} d\varphi \int_{R_{x_0}}^{R} \frac{(z-H)\rho}{(r^2+\rho^2-2r\rho\cos\varphi+(z-H)^2)^{\frac{3}{2}}}\,d\rho,$$

$$w_{22}(r,z) = \frac{1}{4\pi}\int_0^{2\pi} d\varphi \int_{R_{x_0}}^{R} \frac{(z+H)\rho}{(r^2+\rho^2-2r\rho\cos\varphi+(z+H)^2)^{\frac{3}{2}}}\,d\rho,$$

where H is the half thickness of the objective layer, R_0 and R_{x_0} are the radii of the well-hole and the invaded zone, respectively, and R is the radius of the domain under consideration (see Fig. 3.1). By Theorem 4.2 and the superposition principle, $w \overset{\text{def.}}{=} F_3(w_{11}+w_{12}) + F_4(w_{21}+w_{22})$ satisfies

$$\frac{\partial}{\partial r}\left(\sigma r\frac{\partial w}{\partial r}\right) + \frac{\partial}{\partial z}\left(\sigma r\frac{\partial w}{\partial z}\right) = 0 \quad \text{in } \Omega_i \ (i=1,2,3,4), \qquad (4.53)$$

$$\sigma r\frac{\partial w}{\partial n} = 0 \quad \text{on } \bigcup_{k=3}^{6} \Gamma_{2k}, \qquad (4.54)$$

$$\begin{cases} w^+ - w^- = F_k, \\ \left(\dfrac{\partial w}{\partial n}\right)^+ = \left(\dfrac{\partial w}{\partial n}\right)^- \end{cases} \quad \text{on } \gamma_k \ (k=1,\ldots,5). \qquad (4.55)$$

Step 2 Let $v = \tilde{u} - w$. v satisfies

$$\frac{\partial}{\partial r}\left(\sigma r\frac{\partial v}{\partial r}\right) + \frac{\partial}{\partial z}\left(\sigma r\frac{\partial v}{\partial z}\right) = 0 \quad \text{in } \Omega_i \ (i=1,2,3,4), \qquad (4.56)$$

$$v = f \quad \text{on } \Gamma_1, \qquad (4.57)$$

$$\sigma r\frac{\partial v}{\partial n} = g \quad \text{on } \Gamma_2, \qquad (4.58)$$

$$\begin{cases} v^+ - v^- = 0, \\ \left(\sigma r\dfrac{\partial v}{\partial n}\right)^+ - \left(\sigma r\dfrac{\partial v}{\partial n}\right)^- = h_k \end{cases} \quad \text{on } \gamma_k \ (k=1,\ldots,5), \qquad (4.59)$$

where

$$f = -w|_{\Gamma_1},$$

$$g = \sigma r \frac{\partial w}{\partial n}\bigg|_{\Gamma_2},$$

$$h_k = -(\sigma^+ - \sigma^-)\left(r\frac{\partial w}{\partial n}\right)^+\bigg|_{\gamma_k} \quad (k = 1, \ldots, 5).$$

It is easy to see from (4.59) that v is continuous on the interfaces. Therefore, we only need to consider the variational problem corresponding to the problem (4.56)–(4.59): To find $v \in V$, such that for any given $\phi \in V^0$,

$$\sum_{i=1}^{4} \iint_{\Omega_i} \sigma r \left(\frac{\partial v}{\partial r}\frac{\partial \phi}{\partial r} + \frac{\partial v}{\partial z}\frac{\partial \phi}{\partial z}\right) dr dz = \int_{\Gamma_2} g\phi dS - \sum_{k=1}^{5} \int_{\gamma_k} h_k \phi dS, \quad (4.60)$$

where

$$V = \{v \mid v \in H_*^1(\Omega), \ v|_{\Gamma_1} = f\}, \tag{4.61}$$

$$V^0 = \{v \mid v \in H_*^1(\Omega), \ v|_{\Gamma_1} = 0\}. \tag{4.62}$$

Since v is continuous on the interfaces, the variational problem (4.60) can be solved by the common FEM (see Sect. 4.1).

Step 3 Let $u = u_0 + w + v$. We obtain the solution of the original problem (2.4).

In the variational form (4.60), the terms containing g and h_k ($k = 1, \ldots, 5$) possess integral forms which can be treated by the following method in practical calculation.

By Theorem 4.2, it is easy to see that $g \in C^\infty(\Gamma_2)$, $h_k \in C^\infty(\gamma_k)$ ($k = 1, 2, 5$), then we can obtain the desired values by numerical integration. In what follows we only discuss the calculation of h_k ($k = 3, 4$).

Since $\frac{\partial w_{12}}{\partial z}|_{\gamma_3 \cup \gamma_4}$, $\frac{\partial w_{22}}{\partial z}|_{\gamma_3 \cup \gamma_4} \in C^\infty(\gamma_3 \cup \gamma_4)$, $(\frac{\partial w_{12}}{\partial z})^+|_{\gamma_3 \cup \gamma_4}$ and $(\frac{\partial w_{22}}{\partial z})^+|_{\gamma_3 \cup \gamma_4}$ can be easily calculated by numerical integration. For $(\frac{\partial w_{11}}{\partial z})^+|_{\gamma_3}$, $(\frac{\partial w_{21}}{\partial z})^+|_{\gamma_3}$, $(\frac{\partial w_{11}}{\partial z})^+|_{\gamma_4}$ and $(\frac{\partial w_{21}}{\partial z})^+|_{\gamma_4}$, by Lemma 4.3, we take $\delta > 0$ suitable small, and use $\frac{\partial w_{11}}{\partial z}|_{z=H+\delta}$ and $\frac{\partial w_{21}}{\partial z}|_{z=H+\delta}$ to approximate $(\frac{\partial w_{11}}{\partial z})^+|_{z=H}$ and $(\frac{\partial w_{21}}{\partial z})^+|_{z=H}$, respectively, i.e., we take

$$\left(\frac{\partial w_{11}}{\partial z}\right)^+\bigg|_{\gamma_3} \approx \frac{\partial w_{11}}{\partial z}\bigg|_{z=H+\delta}$$

$$= -\frac{1}{4\pi}\int_0^{2\pi} d\varphi \int_{R_0}^{R_{x_0}/2} \frac{(r^2 + \rho^2 - 2r\rho\cos\varphi - 2\delta^2)\rho}{(r^2 + \rho^2 - 2r\rho\cos\varphi + \delta^2)^{\frac{5}{2}}} d\rho.$$

Remark 4.2 Especially, as $\rho_2 = \rho_3 = \rho_4$, by (4.49) we have $h_k = 0$ $(k = 3, 4)$, then the calculation becomes very simple.

Remark 4.3 As $Z \gg H$ and $R \gg R_{x_0}$, by (4.51)–(4.52), f and g are approximately equal to 0 and can be omitted in practical calculation.

References

1. Chen, W., Zuo, L.: One kind of solution methods for the mathematical model of spontaneous potential well-logging. Acta Math. Appl. Sin. **32**(4), 732–745 (2009) (in Chinese)
2. Department of Mathematics of Fudan University: Partial Differential Equations of Mathematical Physics (2nd version). Shanghai Scientific and Technology Press, Shanghai (1961) (in Chinese)
3. Kellogg, O.D.: Foundations of Potential Theory. Springer, Berlin (1929)
4. Li, T., et al.: Advanced Lectures of FEM. Science Press, Beijing (1979) (in Chinese)
5. Li, T., et al.: Applications of the Finite Element Method in Electric Well-Loggings. Oil Industry Press, Beijing (1980) (in Chinese)
6. Li, T., Tan, Y., Peng, Y., Li, H.: Mathematical methods for the SP well-logging. In: Spigler, R. (ed.) Applied and Industrial Mathematics, Venice-1, pp. 343–349. Kluwer Academic, Dordrecht (1991)
7. Li, T., Tan, Y., Peng, Y.: Mathematical model and method for the spontaneous potential well-logging. Eur. J. Appl. Math. **5**, 123–139 (1994)
8. Zhou, Y., Cai, Z.: Convergence of a numerical method in mathematical spontaneous potential well-logging. Eur. J. Appl. Math. **7**(1), 31–41 (1996)
9. Zuo, L., Chen, W.: Double layer potential method for spontaneous potential well-logging. Chin. J. Eng. Math. **25**(6), 1013–1022 (2008) (in Chinese)

Chapter 5
Numerical Simulation

5.1 Uniform Formation

Suppose that the medium is uniform in whole formation, i.e., $\rho_i = \rho_j$ $(i \neq j)$, and that the difference of spontaneous potentials E_4 and E_5 extends until the infinity. In this case there is an exact expression for the spontaneous potential on the well axis as follows [1, 2]:

$$\tilde{u}(0, z) = E_5 + \frac{1}{2}\Delta_A \left(\frac{z+H}{\sqrt{R_0^2 + (z+H)^2}} - \frac{z-H}{\sqrt{R_0^2 + (z-H)^2}} \right)$$
$$+ \frac{1}{2}\Delta_B \left(\frac{z+H}{\sqrt{R_{x_0}^2 + (z+H)^2}} - \frac{z-H}{\sqrt{R_{x_0}^2 + (z-H)^2}} \right). \tag{5.1}$$

We take $H = 2$ m, $R_0 = 0.125$ m, $R_{x_0} = 0.65$ m, $E_1 = 10$ mV, $E_2 = 20$ mV, $E_3 = 100$ mV, $E_4 = 120$ mV, $E_5 = 30$ mV, $R = 1000$ m, $Z = 600$ m and the electrode A_0 is approximately considered as the origin on the symmetric axis. We use the transition zone method and the double layer potential method to do the computation by FEM with about 20000 node points respectively. In the meantime, we use (5.1) to calculate the exact value of spontaneous potential on the well axis. Comparing these values, we find that the error among them on the well axis is quite small (see Fig. 5.1).

Once more we take $E_1 = 20$ mV, $E_2 = 50$ mV, $E_3 = 130$ mV, $E_4 = 140$ mV, $E_5 = 30$ mV, and keep other parameters unchanged. The corresponding results are also very close (see Fig. 5.2).

5.2 Nonuniform Formation

One of the main objectives of the spontaneous potential well-logging is to determine the layer, i.e., to identify the interfaces of the formation.

T. Li et al., *Mathematical Model of Spontaneous Potential Well-Logging and Its Numerical Solutions*, SpringerBriefs in Mathematics,
DOI 10.1007/978-3-642-41425-1_5, © The Author(s) 2014

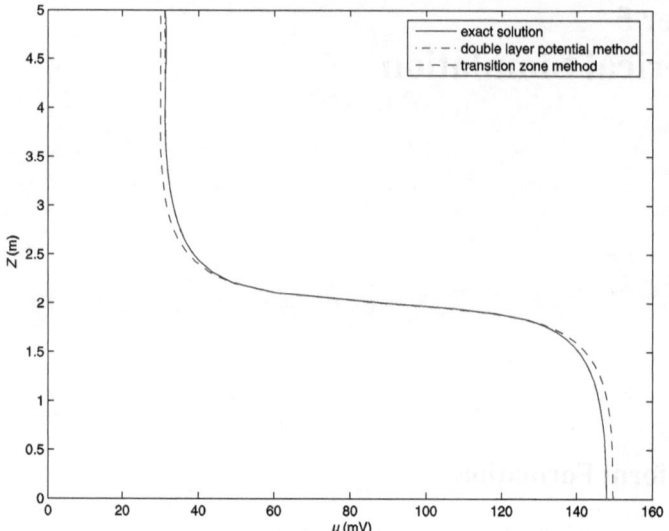

Fig. 5.1   Uniform formation (case (1))

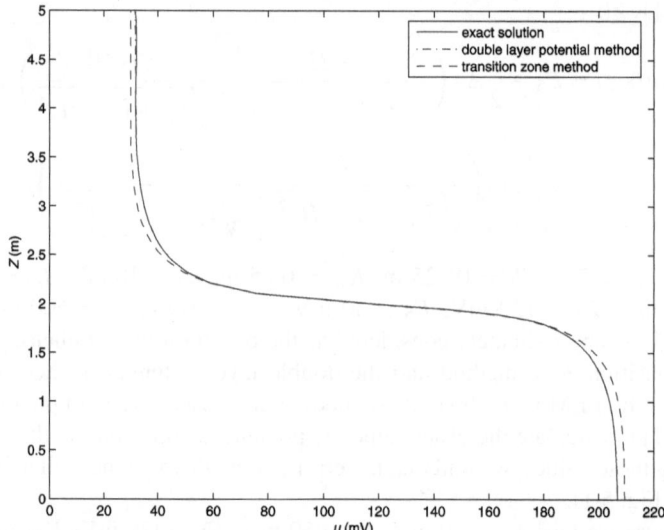

Fig. 5.2   Uniform formation (case (2))

Taking $H = 2$ m, $R_0 = 0.125$ m, $R_{x_0} = 0.65$ m, $E_1 = 10$ mV, $E_2 = 20$ mV, $E_3 = 100$ mV, $E_4 = 120$ mV, $E_5 = 30$ mV, $R = 1000$ m, $Z = 600$ m, we consider the following two cases of nonuniform formation:

$$\rho_1 : \rho_2 : \rho_3 : \rho_4 = 1 : 2 : 2 : 2,$$
$$\rho_1 : \rho_2 : \rho_3 : \rho_4 = 1 : 2 : 3 : 4.$$

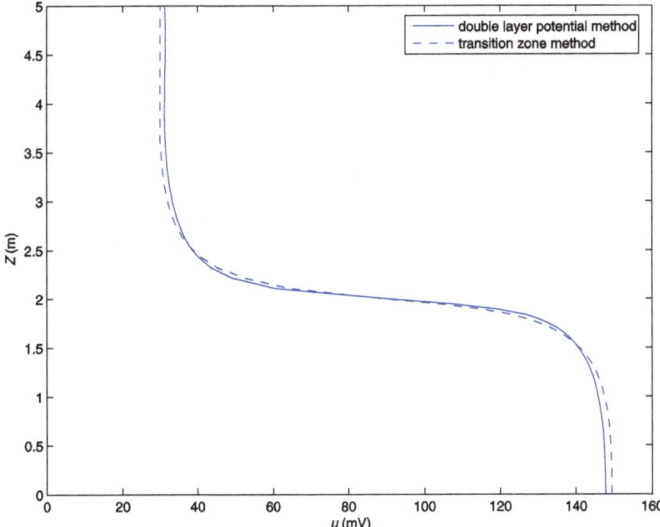

Fig. 5.3 Nonuniform formation $\rho_1 : \rho_2 : \rho_3 : \rho_4 = 1 : 2 : 2 : 2$

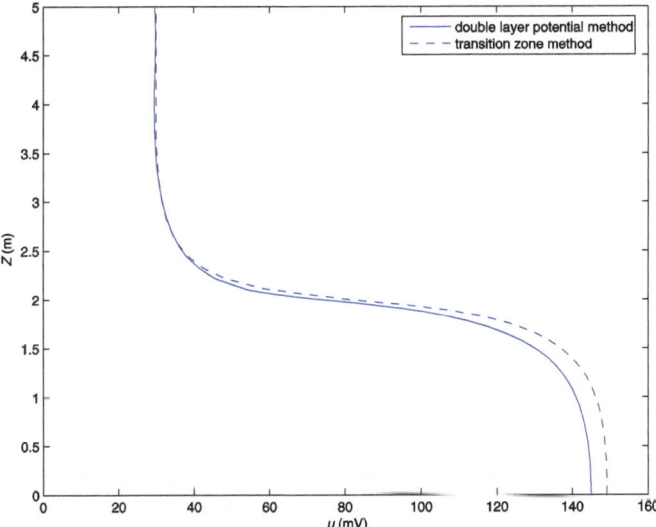

Fig. 5.4 Nonuniform formation $\rho_1 : \rho_2 : \rho_3 : \rho_4 = 1 : 2 : 3 : 4$

By means of the double layer potential method (suppose that the electrode shrinks into a point on the symmetric axis as described in Chap. 3) to do computation and plot the SP well-logging curve (see the dash lines in Figs. 5.3 and 5.4), In the meantime, we use the transition zone method to calculate the SP well-logging curve on

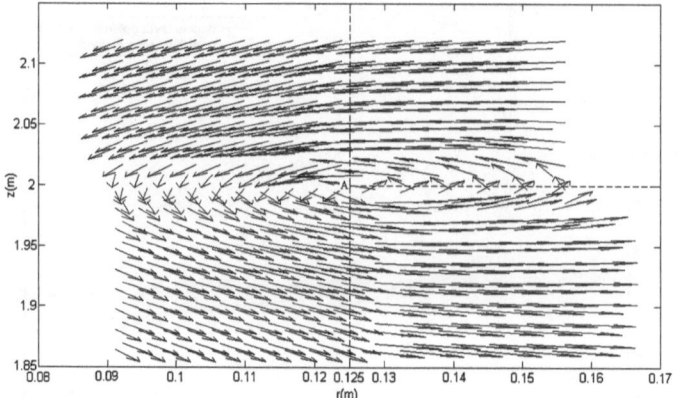

Fig. 5.5 Vectors of electric field for $E_1 = 20$ mV

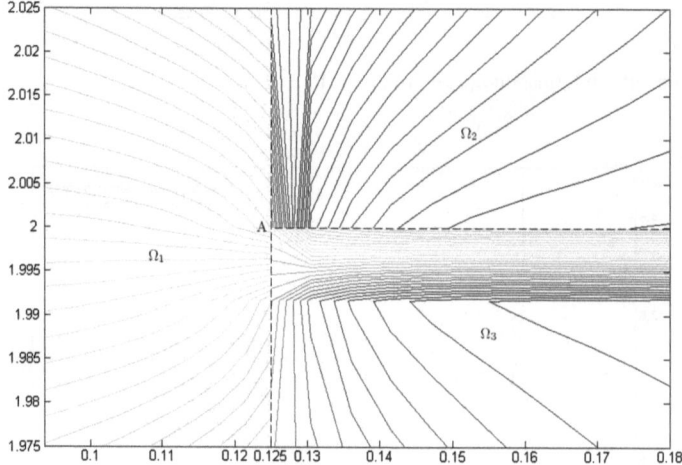

Fig. 5.6 Level curves of potential for $E_1 = 20$ mV

the well axis (see the solid lines in Figs. 5.3 and 5.4). In Figs. 5.3 and 5.4, we see clearly that the results of those two methods are very close.

By numerical simulation we can see the details of the electric field of SP well-logging. We fix the physical parameters and geometric parameters as in the case of Fig. 5.4 except E_1. Take $E_1 = 20$ mV and $E_1 = 250$ mV for the numerical simulation, respectively, and plot the vectors of electric field and the level curves of potential for these two cases, as shown in Figs. 5.5, 5.6 and Figs. 5.7, 5.8, respectively.

From these figures we find two important facts. Firstly, the electric field varies seriously around point A but evenly around point B, since the compatible condition is satisfied at point B but violated at point A in both cases. Secondly, the vectors of electric field around A rotate in different directions, since $\Delta_A \overset{\text{def.}}{=} E_3 - E_1 - E_5$ are

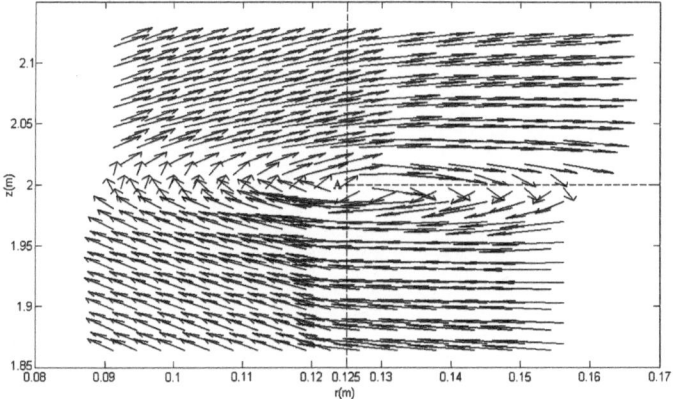

Fig. 5.7 Vectors of electric field for $E_1 = 250$ mV

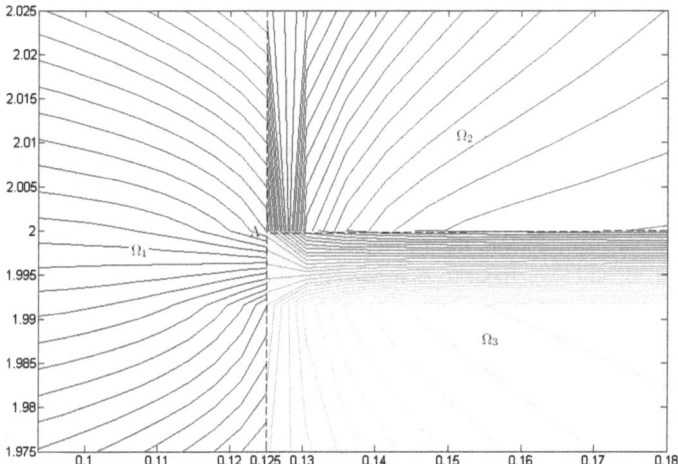

Fig. 5.8 Level curves of potential for $E_1 = 250$ mV

of different signs in these two cases. Indeed, $\Delta_A = E_3 - E_1 - E_5 = 60 > 0$ for the case of $E_1 = 20$ mV, however, $\Delta_A = -130 < 0$ for the case of $E_1 = 250$ mV.

5.3 Impact of Parameters to the Spontaneous Well-Logging Curve

While the well-logging instrument is working, its well-logging response may not exactly reflect the electric parameter information in the objective layer. In fact, the well-logging response is impacted in different degree by the well bore, the invaded zone and the upper and bottom enclosing rocks. Therefore, the resistivity well-

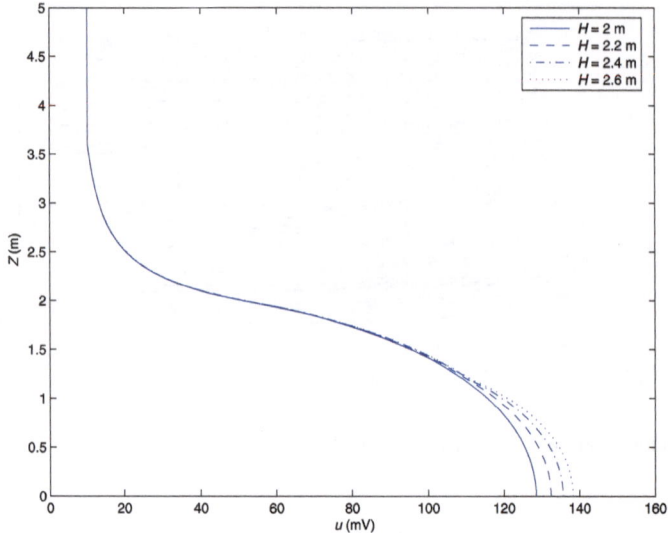

Fig. 5.9 Impact of the thickness H of layer

logging response is called the apparent resistivity which is of difference with the
true resistivity of the objective layer. In order to make the well-logging apparent resistivity response as close as possible to the true resistivity of the objective layer, it
is necessary to do correction, right inversion and interpretation for the well-logging
response. To do this we have to compare the data of well-logging response for a
large number of formation models.

There are many factors which effect the spontaneous potential well-logging response. In what follows, we will do numerical simulation for various thickness of
layer, radius of well bore, radius of invaded zone, resistivity of enclosing rock, resistivity of invaded zone and resistivity of objective layer, and then evaluate the impact
of each factor to the well-logging response.

To facilitate comparison, in all the following numerical experiments the spontaneous potential differences are taken as $E_1 = 20$ mV, $E_2 = 0$ mV, $E_3 = 130$ mV,
$E_4 = 130$ mV, $E_5 = 10$ mV, the size of the formation under consideration is taken
as $R = 1000$ m, $Z = 600$ m. Besides, let $H = 2$ m, $R_0 = 0.125$ m, $R_{x0} = 0.65$ m,
$\rho_1 : \rho_2 : \rho_3 : \rho_4 = 1 : 10 : 50 : 100$ be the basic reference parameter values. In each
time, we only consider the well-logging response varying along with one of the parameters, and other parameters are taken the corresponding values unchanged in this
group of data. In Figs. 5.9–5.14, the SP curves corresponding to this group of basic
parameters are shown by solid lines.

(1) Impact of the thickness H of layer

Figure 5.9 shows the curve of the spontaneous potential response varying along
with the thickness H of layer, where we take

$$H = 2 \text{ m}, 2.2 \text{ m}, 2.4 \text{ m}, 2.6 \text{ m},$$

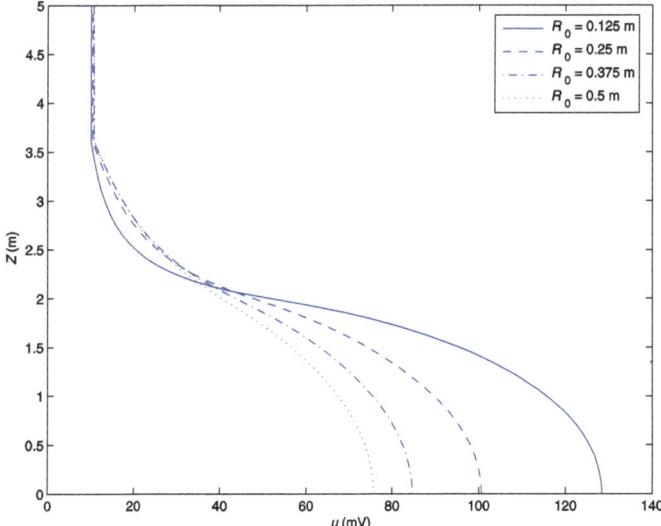

Fig. 5.10   Impact of the radius R_0 of well bore

respectively. It is shown in Fig. 5.9 that for these given data, the varying amplitude of the spontaneous potential curve decreases along with the thickness being thinner.

(2) Impact of the radius R_0 of well bore

Figure 5.10 shows that the curve of spontaneous potential response varying along with the radius R_0 of well bore, where we take

$$R_0 = 0.125 \text{ m}, 0.25 \text{ m}, 0.375 \text{ m}, 0.5 \text{ m},$$

respectively. The result shows that for these given data, the bigger radius of well bore is, the smaller amplitude of spontaneous potential curve will be.

(3) Impact of the radius R_{x_0} of invaded zone

Figure 5.11 shows the curve of spontaneous potential response varying along with the radius R_{x_0} of invaded zone, where we take

$$R_{x_0} = 0.65 \text{ m}, 0.975 \text{ m}, 1.3 \text{ m}, 1.625 \text{ m},$$

respectively. By Fig. 5.11, it is shown that for these given data, the bigger radius R_{x_0} of invaded zone is, i.e., the deeper invasion of the mud to the objective layer is, the smaller amplitude of spontaneous potential curve will be.

(4) Impact of the resistivity ρ_2 of enclosing rock

Figure 5.12 shows the curve of spontaneous potential response varying along with the ratio ρ_2/ρ_1 of surrounding rock resistivity and mud resistivity, where we

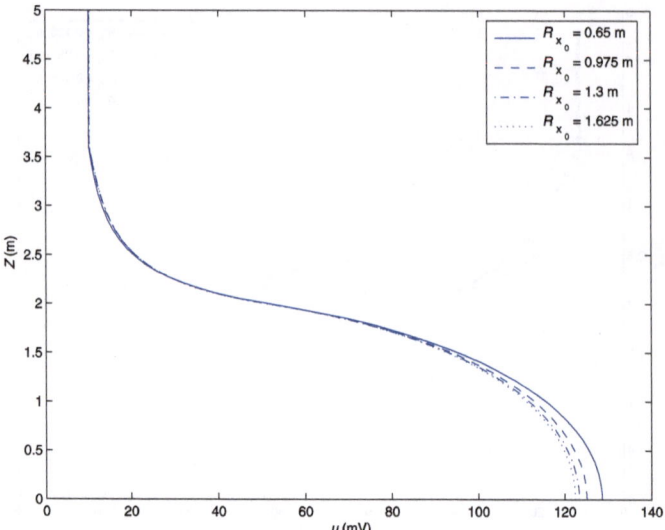

Fig. 5.11 Impact of the radius R_{x_0} of invaded zone

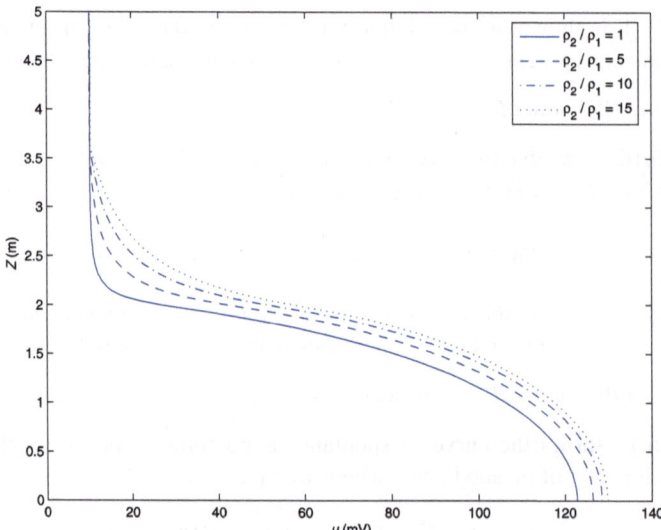

Fig. 5.12 Impact of the resistivity ρ_2 of enclosing rock

take

$$\rho_2/\rho_1 = 1, 5, 10, 15,$$

respectively. By Fig. 5.12, it can be seen that for these given data, the bigger ρ_2/ρ_1 is, the bigger amplitude of spontaneous potential curve will be; on the contrary,

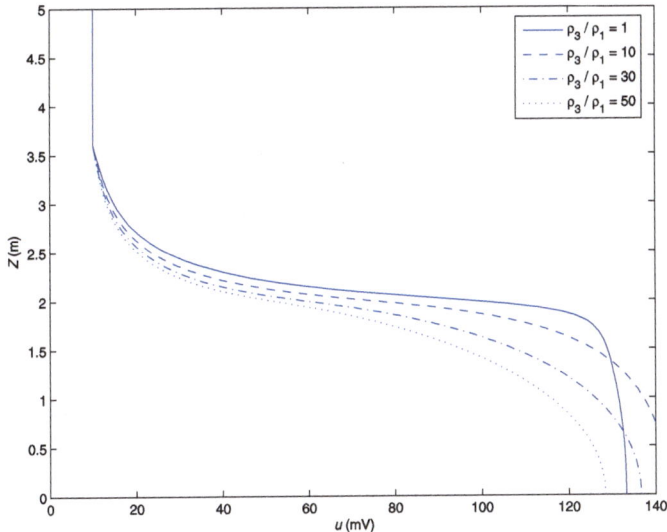

Fig. 5.13 Impact of the resistivity ρ_3 of invaded zone

the smaller ρ_2/ρ_1 is, the smaller amplitude of spontaneous potential curve will be.

(5) Impact of the resistivity ρ_3 of invaded zone

Figure 5.13 shows the curve of spontaneous potential response varying along with the ratio ρ_3/ρ_1 of invaded zone resistivity and mud resistivity, where we take

$$\rho_3/\rho_1 = 1, 10, 30, 50,$$

respectively.

(6) Impact of the resistivity ρ_4 of objective layer

Figure 5.14 shows the curve of spontaneous potential response varying along with the ratio ρ_4/ρ_1 of objective layer resistivity and mud resistivity, where we take

$$\rho_4/\rho_1 = 50, 100, 150, 200,$$

respectively. By Fig. 5.14, it can be seen that the bigger ρ_4/ρ_1 is, the smaller amplitude of spontaneous potential curve will be; on the contrary, the smaller ρ_4/ρ_1 is, the bigger amplitude of spontaneous potential curve will be.

(7) Impact of the resistivity ρ_1 of mud

We now investigate the impact of the resistivity ρ_1 of the mud in the well to the spontaneous potential field. Especially, to meet the practical applications, we examine the behavior of the spontaneous potential curve on the surface of the electrode, i.e., the curve of function $u = u(D/2, z)$, where D is the diameter of the

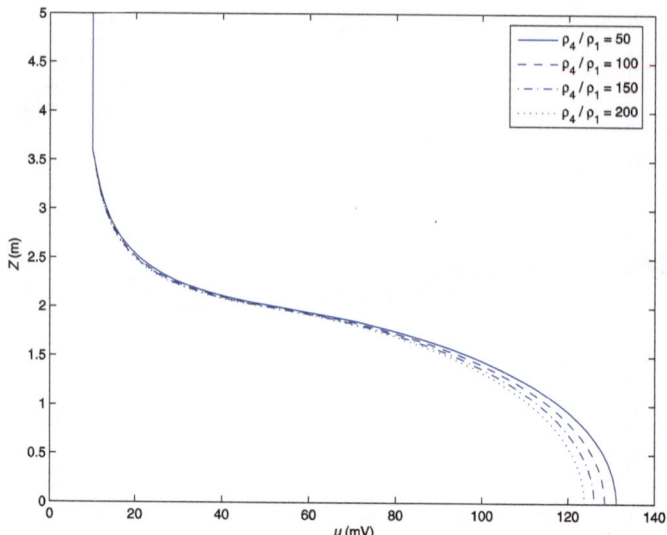

Fig. 5.14 Impact of the resistivity ρ_4 of objective layer

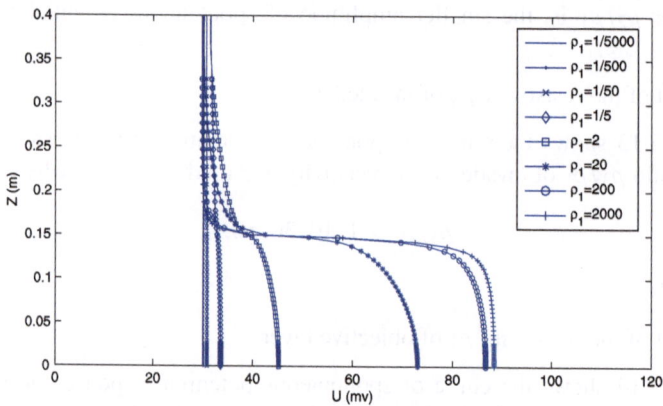

Fig. 5.15 Spontaneous potential curve on the surface of the electrode (case (1))

electrode. Taking $\rho_2 = 10$, $\rho_3 = 50$, $\rho_4 = 100$, and $D = 0.07$ m, we do the numerical simulation for the following two different sets of spontaneous potential differences:

(1) $E_1 = 10$ mV, $E_2 = 20$ mV, $E_3 = 100$ mV, $E_4 = 120$ mV, $E_5 = 30$ mV;
(2) $E_1 = 20$ mV, $E_2 = 0$ mV, $E_3 = 130$ mV, $E_4 = 130$ mV, $E_5 = 10$ mV,

respectively, both of which satisfy the compatible condition $\Delta_B \stackrel{\text{def.}}{=} E_4 - E_3 - E_2 = 0$ at point B, but $\Delta_A \stackrel{\text{def.}}{=} E_3 - E_1 - E_5 \neq 0$ at point A.

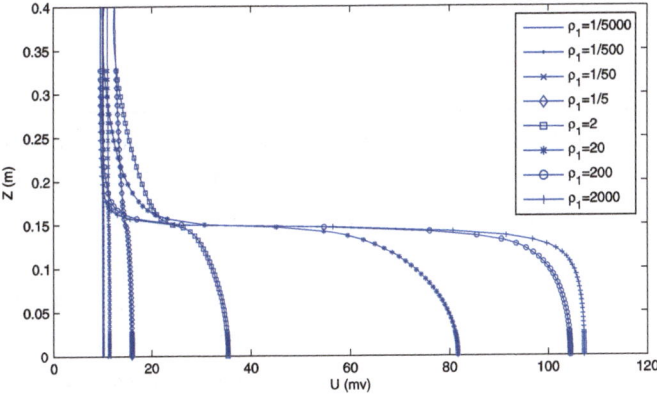

Fig. 5.16 Spontaneous potential curve on the surface of the electrode (case (2))

Let ρ_1 change from small to large, we plot the corresponding spontaneous potential curve on the surface of the electrode as shown in Figs. 5.15 and 5.16 which correspond to case (1) and case (2), respectively.

From these figures we find that the spontaneous potential curve is about a vertical straight line as ρ_1 is small (e.g., $\rho_1 = \frac{1}{5000}, \frac{1}{500}, \frac{1}{50}$), and the spontaneous potential curve behaves a bit rising and falling at the corresponding join point of objective layer and enclosing rock as $\rho_1 = \frac{1}{5}$. This rising and falling becomes higher and higher as long as ρ_1 becomes bigger and bigger. When ρ_1 is very large (e.g., $\rho_1 = 2000$, which means that the well bore is almost isolated) we can find that the variation of the spontaneous potential curve near the inflection point (corresponding to the delimitation of objective layer and enclosing rock) agrees with the value of Δ_A. In case (1), $\Delta_A = 60$, while the maximum and minimum of the spontaneous potential curve near the inflection point are 90 and 30, respectively; In case (2), $\Delta_A = 100$, while the maximum and minimum of the spontaneous potential curve near the inflection point are 110 and 10, respectively. This fact fits very well with the physical intuition.

5.4 Impact of the Structure of Well-Logging Instrument

In Sect. 5.3 we have examined the impact of various physical and geometric parameters to the spontaneous potential well-logging. To simplify the computation, we have not paid much attention to the structure, size and shape of the logging instrument and its components. In this section we will investigate the impact of these factors.

(1) Impact of the diameter of electrode ring

The logging instrument of the spontaneous potential well-logging is mainly composed of a measuring electrode of ring shape embedding on an insolated rubber rod.

Table 5.1 Impact of the diameter of electrode ring (case (1))

$\rho_1 : \rho_2 : \rho_3 : \rho_4$ $= 1 : 10 : 50 : 100$		Diameter D (m) of electrode ring						
		0.05	0.07	0.09	0.11	0.13	0.15	0.17
0.05	U	14.8270	15.1596	15.6200	16.2667	17.2168	18.7395	21.6297
	$\overline{U}$	14.4641	14.5252	14.6077	14.7100	14.8288	14.9576	15.0853
0.10	U	20.3312	20.8812	21.6558	22.7564	24.3880	27.0175	31.9854
	$\overline{U}$	19.7352	19.7936	19.8695	19.9589	20.0564	20.1539	20.2410
0.15	U	25.2526	25.9525	26.9580	28.4051	30.5618	34.0209	40.4196
	$\overline{U}$	24.5064	24.5482	24.6028	24.6660	24.7340	24.8015	24.8613
0.20	U	29.7262	30.5563	31.7605	33.4994	36.0834	40.1841	47.5930
	$\overline{U}$	28.8533	28.8853	28.9262	28.9740	29.0256	29.0769	29.1224
0.25	U	33.8543	34.7979	36.1714	38.1527	41.0799	45.6703	53.7703
	$\overline{U}$	32.8707	32.8965	32.9296	32.9682	33.0100	33.0517	33.0888

Half thickness H (m) of objective layer (row label at left)

The electrode is a metal ring, i.e., a metal cylindric surface. In practical applications, this rings possesses several different sizes. We now examine the impact of the diameter of electrode ring.

Taking the isolated rod with half length 6 m, and $R = 50$ m, $Z = 50$ m, $R_0 = 0.1016$ m, $R_{x0} = 0.65$ m, $E_1 = 20$ mV, $E_2 = 0$ mV, $E_3 = 130$ mV, $E_4 = 130$ mV, $E_5 = 10$ mV, we compute the spontaneous potential field for the following two cases of different formation resistivity:

(1) $\rho_1 : \rho_2 : \rho_3 : \rho_4 = 1 : 10 : 50 : 100$;
(2) $\rho_1 : \rho_2 : \rho_3 : \rho_4 = 1 : 1000 : 50 : 100$,

respectively.

In Tables 5.1 and 5.2 we show the result of electric potential for different diameters in several different half-thicknesses H of objective layer for cases (1) and (2), respectively, where U is the spontaneous potential on the electrode surface (i.e., the potential value on $r = D/2$), and $\overline{U}$ is the corresponding value of spontaneous potential at the origin, i.e., the value obtained by neglecting the insolated rod.

From Table 5.1 and Table 5.2 we find that fixing other factors, the spontaneous potential on the electrode surface becomes bigger as the diameter becomes bigger. Moreover, the spontaneous potential on the electrode surface becomes smaller as the thickness of objective layer becomes smaller. Meanwhile, the computation shows that the effect of the isolated rubber rod can not be ignored.

(2) Impact of the length of isolated rod

As just mentioned above, the impact of the isolated rod is significant. Now we do some simulation to analyse quantitatively the impact of the rod size.

Taking $R = 50$ m, $Z = 50$ m, $R_0 = 0.1016$ m, $R_{x0} = 0.65$ m, $\rho_1 : \rho_2 : \rho_3 : \rho_4 = 1 : 10 : 50 : 100$, $E_1 = 20$ mV, $E_2 = 0$ mV, $E_3 = 130$ mV, $E_4 = 130$ mV,

Table 5.2 Impact of the diameter of electrode ring (case (2))

$\rho_1 : \rho_2 : \rho_3 : \rho_4$ $= 1 : 1000 : 50 : 100$		Diameter D (m) of electrode ring						
		0.05	0.07	0.09	0.11	0.13	0.15	0.17
0.05	U	30.0097	30.7413	31.7968	33.2778	35.3732	38.4669	43.5187
	$\overline{U}$	30.4393	30.4546	30.4752	30.5007	30.5304	30.5632	30.5969
0.10	U	42.8103	43.8204	45.2657	47.2630	50.0237	53.9610	60.0519
	$\overline{U}$	43.5520	43.5666	43.5859	43.6090	43.6349	43.6621	43.6881
0.15	U	51.7891	52.9126	54.5137	56.7053	59.6894	63.8507	70.0688
	$\overline{U}$	52.7475	52.7591	52.7742	52.7923	52.8122	52.8329	52.8523
0.20	U	58.5158	59.6826	61.3413	63.5963	66.6330	70.7990	76.8716
	$\overline{U}$	59.6262	59.6356	59.6478	59.6623	59.6783	59.6948	59.7106
0.25	U	63.7888	64.9649	66.6342	68.8919	71.9063	75.9900	81.8298
	$\overline{U}$	65.0075	65.0154	65.0256	65.0377	65.0510	65.0646	65.0774

Half thickness H (m) of objective layer

$E_5 = 10$ mV, we compute the spontaneous potential field for different lengths of the rubber rod in different half thicknesses of the objective layer for the following 3 cases of electrode diameter:

(1) $D = 0.07$ m;
(2) $D = 0.09$ m;
(3) $D = 0.18$ m,

respectively.

We use $H_J = 0$ to represent the half length of the rod. Tables 5.3, 5.4 and 5.5 list the corresponding potential value on $r = D/2$, i.e., on the surface of the electrode, where $H_J = 0$ means the isolated rod being neglected.

In these tables we find that fixing other factors, the spontaneous potential on the electrode surface becomes bigger as the isolated rod becomes longer. The variation is more sensitive when the rod is shorter, while the change becomes slower as the rod is longer. Particularly, the spontaneous potential on the electrode surface keeps almost unchanged as the half length of the rod is longer than 1.98 m (see the bold-faced digits in the tables).

(3) Impact of the height of electrode

We now investigate the impact of the height of electrode in the case of thin objective layer. Let the half length of the isolated rod be 6 m. Taking $R = 50$ m, $Z = 50$ m, $R_0 = 0.1016$ m, $R_{x_0} = 0.65$ m, $\rho_1 : \rho_2 : \rho_3 : \rho_4 = 1 : 10 : 50 : 100$, $E_1 = 20$ mV, $E_2 = 0$ mV, $E_3 = 130$ mV, $E_4 = 130$ mV, $E_5 = 10$ mV, we compute the spontaneous potential on the electrode surface for different heights of the electrode in different thicknesses of the objective layer for the following 2 cases of electrode diameter:

Table 5.3 Impact of the rod length (case (1))

$D = 0.07$ m	Half thickness H (m) of the objective layer							
Half length H_J (m) of the rod	0.0090	0.0360	0.0630	0.0900	0.1170	0.1440	0.1710	0.1980
0	10.6748	13.3019	16.1492	18.8955	21.4628	23.8408	26.0106	27.8438
0.009	10.7134	13.4612	16.3609	19.0997	21.6395	23.9904	26.1388	27.9560
0.018	10.7232	13.5047	16.4242	19.1643	21.6967	24.0391	26.1806	27.9926
0.027	10.7298	13.5359	16.4742	19.2188	21.7464	24.0819	26.2173	28.0246
0.036	10.7349	13.5609	16.5175	19.2696	21.7947	24.1240	26.2535	28.0562
0.045	10.7389	13.5815	16.5561	19.3185	21.8434	24.1674	26.2910	28.0888
0.054	10.7423	13.5988	16.5904	19.3657	21.8931	24.2127	26.3305	28.1231
0.063	10.7450	13.6135	16.6209	19.4107	21.9437	24.2606	26.3726	28.1598
0.072	10.7474	13.6261	16.6477	19.4529	21.9946	24.3108	26.4177	28.1991
0.081	10.7494	13.6368	16.6713	19.4919	22.0451	24.3633	26.4658	28.2413
0.090	10.7512	13.6462	16.6920	19.5274	22.0941	24.4174	26.5170	28.2865
0.108	10.7541	13.6615	16.7261	19.5878	22.1843	24.5269	26.6273	28.3861
0.126	10.7563	13.6734	16.7526	19.6356	22.2603	24.6308	26.7439	28.4965
0.144	10.7581	13.6827	16.7732	19.6729	22.3212	24.7208	26.8573	28.6117
0.162	10.7594	13.6895	16.7884	19.7002	22.3662	24.7899	26.9520	28.7161
0.180	10.7601	13.6930	16.7960	19.7138	22.3885	24.8245	27.0014	28.7740
0.360	10.7642	13.7151	16.8442	19.7974	22.5196	25.0190	27.2799	29.1594
0.540	10.7650	13.7200	16.8552	19.8166	22.5493	25.0620	27.3397	29.2402
0.720	**10.7652**	13.7213	16.8582	19.8219	22.5575	25.0739	27.3563	29.2623
0.900	**10.7652**	13.7216	16.8591	19.8235	22.5602	25.0778	27.3616	29.2694
1.080	**10.7652**	**13.7218**	16.8594	19.8241	22.5611	25.0792	27.3635	29.2719
1.260	**10.7652**	**13.7218**	16.8595	19.8243	22.5615	25.0797	27.3642	29.2729
1.440	**10.7652**	**13.7218**	**16.8596**	19.8244	22.5616	25.0799	27.3645	29.2733
1.620	**10.7652**	**13.7218**	**16.8596**	**19.8245**	**22.5617**	**25.0800**	27.3646	29.2734
1.800	**10.7652**	**13.7218**	**16.8596**	**19.8245**	**22.5617**	**25.0800**	**27.3647**	**29.2735**
1.980	**10.7652**	**13.7218**	**16.8596**	**19.8245**	**22.5617**	**25.0800**	**27.3647**	**29.2735**
2.160	**10.7652**	**13.7218**	**16.8596**	**19.8245**	**22.5617**	**25.0800**	**27.3647**	**29.2735**
⋮								
9.000								

(1) $D = 0.07$ m;
(2) $D = 0.17$ m,

respectively.

Let ε be the height of electrode. We denote the spontaneous potential on the electrode surface by U_ε and $\overline{U}_\varepsilon$ which correspond to the situations with or without considering the isolated rod, respectively. As $\varepsilon = 0$, the cylinder surface electrode

Table 5.4 Impact of the rod length (case (2))

$D = 0.09$ m	Half thickness H (m) of the objective layer							
Half length H_J (m) of the rod	0.0090	0.0360	0.0630	0.0900	0.1170	0.1440	0.1710	0.1980
0	10.6891	13.3615	16.2297	18.9733	21.5300	23.8976	26.0593	27.8864
0.009	10.7522	13.6209	16.5670	19.2925	21.8034	24.1284	26.2570	28.0597
0.018	10.7675	13.6899	16.6674	19.3933	21.8917	24.2032	26.3211	28.1158
0.027	10.7780	13.7411	16.7507	19.4833	21.9729	24.2728	26.3807	28.1678
0.036	10.7861	13.7824	16.8250	19.5707	22.0550	24.3439	26.4417	28.2209
0.045	10.7926	13.8167	16.8919	19.6569	22.1401	24.4190	26.5064	28.2772
0.054	10.7980	13.8455	16.9517	19.7413	22.2288	24.4993	26.5759	28.3376
0.063	10.8025	13.8698	17.0044	19.8221	22.3203	24.5850	26.6510	28.4028
0.072	10.8064	13.8906	17.0506	19.8979	22.4132	24.6761	26.7320	28.4733
0.081	10.8097	13.9084	17.0908	19.9673	22.5055	24.7719	26.8192	28.5494
0.090	10.8125	13.9238	17.1258	20.0297	22.5949	24.8711	26.9124	28.6314
0.108	10.8172	13.9490	17.1829	20.1341	22.7568	25.0721	27.1144	28.8127
0.126	10.8209	13.9685	17.2268	20.2149	22.8898	25.2602	27.3282	29.0144
0.144	10.8237	13.9835	17.2606	20.2766	22.9930	25.4184	27.5332	29.2242
0.162	10.8258	13.9944	17.2847	20.3204	23.0660	25.5337	27.6978	29.4099
0.180	10.8268	13.9996	17.2964	20.3414	23.1007	25.5887	27.7794	29.5093
0.360	10.8336	14.0360	17.3761	20.4807	23.3204	25.9162	28.2516	30.1690
0.540	10.8348	14.0437	17.3936	20.5114	23.3683	25.9861	28.3492	30.3016
0.720	10.8350	14.0456	17.3981	20.5195	23.3811	26.0048	28.3753	30.3366
0.900	**10.8351**	14.0461	17.3994	20.5220	23.3851	26.0106	28.3833	30.3474
1.080	**10.8351**	14.0462	17.3999	20.5229	23.3865	26.0126	28.3861	30.3511
1.260	**10.8351**	**14.0463**	17.4000	20.5232	23.3870	26.0134	28.3872	30.3525
1.440	**10.8351**	**14.0463**	**17.4001**	20.5233	23.3871	26.0137	28.3876	30.3531
1.620	**10.8351**	**14.0463**	**17.4001**	**20.5234**	23.3872	26.0138	28.3877	30.3533
1.800	**10.8351**	**14.0463**	**17.4001**	**20.5234**	**23.3873**	26.0138	**28.3878**	**30.3534**
1.980	**10.8351**	**14.0463**	**17.4001**	**20.5234**	**23.3873**	**26.0139**	**28.3878**	**30.3534**
2.160	**10.8351**	**14.0463**	**17.4001**	**20.5234**	**23.3873**	**26.0139**	**28.3878**	**30.3534**
⋮								
9.000								

reduces into a circle and the corresponding spontaneous potential on the electrode surface are denoted by U_0 and $\overline{U}_0$.

Tables 5.6 and 5.7 list the numerical results which show that as $\varepsilon \to 0$, the values U_ε and $\overline{U}_\varepsilon$ of spontaneous potential on the electrode surface tend to U_0 and $\overline{U}_0$, respectively, even for the case with very thin objective layer. It validates the limiting behavior mentioned in Chap. 3 once again. Moreover, even with the existence of the isolated rod, the conclusion of limiting behavior is still valid.

Table 5.5 Impact of the rod length (case 3)

$D = 0.18$ m	Half thickness H (m) of the objective layer							
	0.0090	0.0360	0.0630	0.0900	0.1170	0.1440	0.1710	0.1980
0	10.7537	13.7540	16.7235	19.4058	21.8885	24.1974	26.3161	28.1122
0.009	11.2711	16.2351	19.1327	21.3255	23.4419	25.4955	27.4333	29.0989
0.018	11.4008	17.0893	20.2518	22.2089	24.1482	26.0827	27.9377	29.5440
0.027	11.4904	17.7033	21.4687	23.2176	24.9523	26.7479	28.5069	30.0452
0.036	11.5575	18.1536	22.6936	24.3662	25.8690	27.5022	29.1497	30.6093
0.045	11.6093	18.4925	23.7235	25.6449	26.9024	28.3479	29.8669	31.2366
0.054	11.6501	18.7536	24.5071	27.0129	28.0547	29.2870	30.6590	31.9268
0.063	11.6827	18.9586	25.1017	28.3622	29.3246	30.3218	31.5269	32.6796
0.072	11.7090	19.1219	25.5614	29.4948	30.7003	31.4550	32.4719	33.4952
0.081	11.7304	19.2538	25.9238	30.3683	32.1423	32.6884	33.4961	34.3742
0.090	11.7480	19.3611	26.2139	31.0420	33.5471	34.0208	34.6019	35.3178
0.108	11.7746	19.5222	26.6413	31.9932	35.6433	36.9100	37.0674	37.4044
0.126	11.7931	19.6333	26.9311	32.6140	36.9236	39.5197	39.8572	39.7669
0.144	11.8060	19.7104	27.1302	33.0312	37.7425	41.1716	42.7150	42.3919
0.162	11.8145	19.7618	27.2625	33.3046	38.2647	42.1654	44.7281	45.0945
0.180	11.8187	19.7872	27.3276	33.4381	38.5159	42.6294	45.6264	46.8456
0.360	11.8331	19.8836	27.5771	33.9335	39.3876	44.0686	47.9229	50.4414
0.540	11.8331	19.8909	27.6017	33.9858	39.4806	44.2190	48.1527	50.7810
0.720	11.8325	19.8907	27.6040	33.9926	39.4937	44.2410	48.1864	50.8305
0.900	11.8323	19.8903	27.6041	33.9935	39.4961	44.2452	48.1930	50.8403
1.080	**11.8321**	19.8901	27.6040	**33.9937**	39.4965	44.2461	48.1946	50.8427
1.260	**11.8321**	**19.8900**	27.6040	**33.9937**	**39.4967**	44.2464	48.1951	50.8434
1.440	**11.8321**	**19.8900**	**27.6039**	**33.9937**	**39.4967**	**44.2465**	48.1952	50.8436
1.620	**11.8321**	**19.8900**	**27.6039**	**33.9937**	**39.4967**	**44.2465**	**48.1953**	**50.8437**
1.800	**11.8321**	**19.8900**	**27.6039**	**33.9937**	**39.4967**	**44.2465**	**48.1953**	**50.8437**
1.980	**11.8321**	**19.8900**	**27.6039**	**33.9937**	**39.4967**	**44.2465**	**48.1953**	**50.8437**
2.160	**11.8321**	**19.8900**	**27.6039**	**33.9937**	**39.4967**	**44.2465**	**48.1953**	**50.8437**
⋮								
9.000								

Half length H_J (m) of the rod (left axis label)

(4) Impact of installing the spontaneous potential well-logging instrument on the lateral well-logging instrument

In practical well-loggings in the oil field, the spontaneous potential well-logging and the lateral well-logging are often combined to use, and the spontaneous potential well-logging instrument is installed on the upper position of the lateral well-logging tool (see Fig. 5.18). In Fig. 5.17, A is the spontaneous potential measuring electrode ring with diameter 30 mm and length 250 mm; B is a section of cable with insulating

Table 5.6 Impact of the electrode height (case (1))

$D = 0.07$ m		Height of electrode ε (m)								
		0	0.014	0.034	0.054	0.074	0.094	0.114	0.134	0.154
Half thickness H (m) of objective layer	0.05 U_ε	15.7584	15.0986	15.1372	15.1183	15.0489	14.9365	14.7796	14.6045	14.4290
	$\overline{U}_\varepsilon$	15.1963	14.5017	14.5418	14.5252	14.4631	14.3641	14.2291	14.0872	13.9403
	0.10 U_ε	21.3082	20.8069	20.8288	20.8119	20.7567	20.6614	20.5236	20.3414	20.1141
	$\overline{U}_\varepsilon$	20.3025	19.8000	19.8182	19.7937	19.7301	19.6282	19.4868	19.3066	19.0888
	0.15 U_ε	26.1771	25.8631	25.8769	25.8638	25.8252	25.7600	25.6662	25.5414	25.3827
	$\overline{U}_\varepsilon$	24.8787	24.5571	24.5682	24.5482	24.5018	24.4276	24.3247	24.1913	24.0257
	0.20 U_ε	30.6251	30.4588	30.4557	30.4548	30.4107	30.3713	30.2871	30.2049	30.0745
	$\overline{U}_\varepsilon$	29.0715	28.8950	28.8915	28.8853	28.8367	28.7898	28.7000	28.6098	28.4742
	0.25 U_ε	34.7069	34.7085	34.7084	34.6885	34.6534	34.6032	34.5365	34.4865	34.3832
	$\overline{U}_\varepsilon$	32.9320	32.9190	32.9187	32.8965	32.8581	32.8040	32.7330	32.6757	32.5686

Table 5.7 Impact of the electrode height (case (2))

$D = 0.17$ m		Height of electrode ε (m)								
		0	0.014	0.034	0.054	0.074	0.094	0.114	0.134	0.154
Half thickness H (m) of objective layer	0.05 U_ε	22.2110	21.8646	21.8725	21.6023	20.9936	20.0063	18.7255	17.5329	16.7308
	$\overline{U}_\varepsilon$	16.1929	15.1882	15.1913	15.0853	14.8681	14.5403	14.1394	13.7906	13.5194
	0.10 U_ε	32.7218	32.0610	32.0657	31.9422	31.6795	31.2698	30.7025	29.9582	29.0025
	$\overline{U}_\varepsilon$	21.0770	20.2766	20.2879	20.2410	20.1394	19.9818	19.7651	19.4834	19.1274
	0.15 U_ε	40.8098	40.4740	40.4644	40.3701	40.1888	39.9193	39.5602	39.1091	38.5614
	$\overline{U}_\varepsilon$	25.4210	24.8902	24.8963	24.8613	24.7890	24.6801	24.5343	24.3507	24.1279
	0.20 U_ε	47.8075	47.6284	47.5898	47.5420	47.3755	47.2034	46.9134	46.6185	46.2038
	$\overline{U}_\varepsilon$	29.4874	29.1496	29.1418	29.1225	29.0527	28.9772	28.8523	28.7219	28.5416
	0.25 U_ε	53.8297	53.8186	53.7969	53.7202	53.5956	53.4253	53.2093	53.0236	52.7145
	$\overline{U}_\varepsilon$	33.2713	33.1263	33.1220	33.0888	33.0323	32.9538	32.8531	32.7615	32.6165

surface, which can be regarded as an insulator with diameter 30 mm and length 4.5 m; C is the conductor with diameter 90 mm and three different lengths 5 m, 10 m, 15 m, respectively; D is an insulator with diameter 90 mm and length 800 mm; E denotes the dual lateral log tool and acoustic instruments with diameter 90 mm, and its length information is shown in Fig. 5.18.

When we do the lateral well-logging, the spontaneous potential well-logging instrument does not work; while, when we do the spontaneous potential well-logging, the lateral well-logging instrument does not work. Note that a lateral well-logging tool is composed of many electrodes with different sizes (including an emitting elec-

Fig. 5.17 Combination of
the SP well-logging
instrument and the lateral
well-logging instrument

trode, a number of shield electrodes and measuring electrodes), moreover, a cylindrical conductor C (which can be regarded as an electrode without emitting current) and two cylindrical insulating cables B and D are inserted to separate these two kinds of well-logging instruments. When we do the spontaneous potential well-logging, although those electrodes do not work, i.e., they do not emit current, however, due to their existence they certainly occupy a part of the domain of the original spontaneous potential field near the well axis, and the original insulating condition essentially satisfied on a part of the well axis should be replaced by the homogeneous equi-potential surface boundary condition on those electrodes Γ_k $(k = 1, \ldots, l)$:

$$u|_{\Gamma_k} = c_k \quad \text{(constant to be determined),} \tag{5.2}$$

$$\int_{\Gamma_k} \sigma r \frac{\partial u}{\partial n} \, dS = 0 \tag{5.3}$$

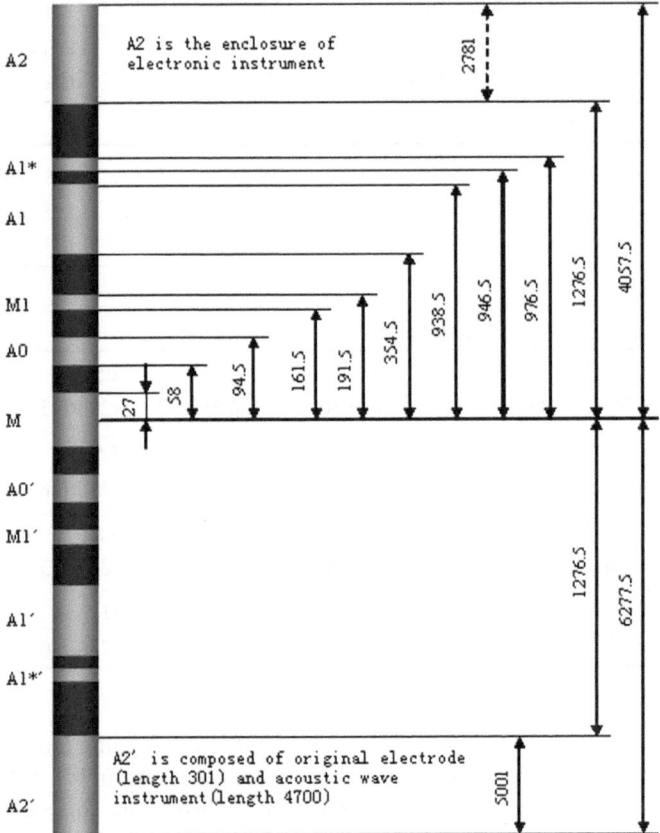

Fig. 5.18 Data of the lateral well-logging tool

and by the insulating boundary conditions

$$\sigma r \frac{\partial u}{\partial n}\bigg|_{\widetilde{\Gamma}_h} = 0 \qquad (5.4)$$

on some insulating boundaries $\widetilde{\Gamma}_h$ ($h = 1, \ldots, m$). Then, the original equi-potential surface boundary value problem (1.5)–(1.13) for the spontaneous potential well-logging should be replaced by the equi-potential surface boundary value problem (1.5)–(1.13) and (5.2)–(5.4), which possesses a different solution domain and some different boundary conditions from the original problem. Thus, the whole spontaneous potential field should be accordingly changed, especially the spontaneous potential U on the measuring electrode Γ_0, which is the main object concerned by the spontaneous potential well-logging, should be correspondingly changed. A natural question is if these changes will bring serious influence to the value of the spontaneous potential U on the measuring electrode.

Table 5.8 Impact of the electrodes of the lateral well-logging tool to the SP well-logging

C	First group $U = 148.7506$		Second group $U = 126.5869$		Third group $U = 207.9947$		Fourth group $U = 171.2460$	
	$\tilde{U}$	RE	$\tilde{U}$	RE	$\tilde{U}$	RE	$\tilde{U}$	RE
5	148.7953	0.03 %	127.7263	0.9 %	208.0677	0.04 %	173.1444	1.11 %
10	148.7953	0.03 %	127.7262	0.9 %	208.0677	0.04 %	173.1444	1.11 %
15	148.7953	0.03 %	127.7262	0.9 %	208.0677	0.04 %	173.1444	1.11 %

Table 5.9 Impact of the length of cable B

B	0.5	1	1.5	2	2.5	3
$\tilde{U}$	55.0730	86.1598	110.5930	126.2181	128.9323	129.0688
B	3.5	4	4.5	5	5.5	6
$\tilde{U}$	128.4146	128.3631	*127.7263*	127.6038	127.2428	**126.8082**
B	6.5	7	7.5	8	8.5	9
$\tilde{U}$	**126.2426**	125.4525	125.3500	125.1463	124.9613	124.7798
B	9.5	10	10.5	11	11.5	12
$\tilde{U}$	124.5876	124.3707	124.1156	123.8084	123.4346	122.9778

To test the impact of the electrodes of the lateral well-logging tool to the spontaneous potential well-logging, we use the tool data as shown in Fig. 5.18 and take the half thickness of layer $H = 2$ m, the radius of well bore $R_0 = 0.125$ m, the radius of invaded zone $R_0 = 0.125$ m, and the radial depth and the vertical height $R = 1000$ m and $Z = 600$ m, respectively, to do numerical computation by the transition zone method, Denoting the value of the spontaneous potential on the measuring electrode Γ_0 without and with inserting electrodes of the lateral well-logging tool by U and $\tilde{U}$, respectively, we compute them by using the following data, and show the results in Table 5.8.

(1) $\rho_1 : \rho_2 : \rho_3 : \rho_4 = 1 : 2 : 3 : 4$, $E_1 = 10$ mV, $E_2 = 20$ mV, $E_3 = 100$ mV, $E_4 = 120$ mV, $E_5 = 30$ mV;
(2) $\rho_1 : \rho_2 : \rho_3 : \rho_4 = 1 : 10 : 50 : 100$, $E_1 = 10$ mV, $E_2 = 20$ mV, $E_3 = 100$ mV, $E_4 = 120$ mV, $E_5 = 30$ mV;
(3) $\rho_1 : \rho_2 : \rho_3 : \rho_4 = 1 : 2 : 3 : 4$, $E_1 = 20$ mV, $E_2 = 50$ mV, $E_3 = 130$ mV, $E_4 = 140$ mV, $E_5 = 10$ mV;
(4) $\rho_1 : \rho_2 : \rho_3 : \rho_4 = 1 : 10 : 50 : 100$, $E_1 = 20$ mV, $E_2 = 50$ mV, $E_3 = 130$ mV, $E_4 = 140$ mV, $E_5 = 10$ mV.

For the length of C taken as 5 m, 10 m and 15 m, we do computation respectively, the results show that the impact of inserting the electrodes of the lateral well-logging instrument to the spontaneous potential well-logging is actually very small (see Table 5.8, in which RE means the relative error).

Below we will focus on the impact of the length of cable B which separates two groups of electrodes and is of the insulating surface, to the spontaneous potential U on the measuring electrode. We take $E_1 = 10$ mV, $E_2 = 20$ mV, $E_3 = 100$ mV, $E_4 = 120$ mV, $E_5 = 30$ mV, $\rho_1 : \rho_2 : \rho_3 : \rho_4 = 1 : 10 : 50 : 100$ to do the computation. In this case, without inserting the electrodes of lateral well-logging tool, the value of spontaneous potential U on the measuring electrode is 126.5869. The results can be found in Table 5.9.

From Table 5.9, we find that the insulating cable B can not be too long and especially can not be too short, otherwise it will (may be seriously) effect the result of the spontaneous potential well-logging. For the instrument given by Fig. 5.14, the length of B is taken as 4.5 m and the relative error $\frac{\tilde{U}-U}{U}$ is about 0.9 %, which meets the need of engineering. A better choice of the length of B should be 6 m or 6.5 m, for which the relative errors are 0.18 % and 0.27 %, respectively.

References

1. Li, H.: Exact solutions to the spontaneous potential well-logging equation in uniform formation. Cnin. Ann. Math. **17A**(1):87–96 (1996) (in Chinese)
2. Peng, Y.: Mathematical method of spontaneous potential well-logging. Appl. Math. Comput. Math. **2**(2), 35–43 (1988) (in Chinese)